THE END OF DAYS

ARMAGEDDON AND PROPHECIES
OF THE RETURN

ALSO BY ZECHARIA SITCHIN

THE EARTH CHRONICLES

COMPANION BOOKS

ZECHARIA SITCHIN

THE END OF DAYS

ARMAGEDDON AND PROPHECIES OF THE RETURN

The 7th and Concluding Book of *The Earth Chronicles*

wm

WILLIAM MORROW

An Imprint of HarperCollins*Publishers*

HarperCollins books may be purchased for educational, business, or sales
promotional use. For information please write: Special Markets Department, HarperCollins Publishers, 10 East 53rd Street, New York, NY
10022.

Designed by Renato Stanisic

Library of Congress Cataloging-in-Publication Data has been applied for.

ISBN: 978-0-06-123823-9
ISBN-10: 0-06-123823-6

08 09 10 11 WBC/RRD 10 9 8 7 6

DEDICATED

TO MY BROTHER

DR. AMNON SITCHIN,

WHOSE AEROSPACE EXPERTISE

WAS INVALUABLE AT ALL TIMES

Contents

Preface:
THE PAST, THE FUTURE

"When will they return?"

I have been asked this question countless times by people who have read my books, the "they" being the Anunnaki—the extra-terrestrials who had come to Earth from their planet Nibiru and were revered in antiquity as gods. Will it be when Nibiru in its elongated orbit returns to our vicinity, and what will happen then? Will there be darkness at noon and the Earth shall shatter? Will it be Peace on Earth, or Armageddon? A Millennium of trouble and tribulations, or a messianic Second Coming? Will it happen in 2012, or later, or not at all?

These are profound questions that combine people's deepest hopes and anxieties with religious beliefs and expectations, questions compounded by current events: wars in lands where the entwined affairs of gods and men began; the threats of nuclear holocausts; the alarming ferocity of natural disasters. They are questions that I dared not answer all these years—but now are questions the answers to which cannot—must not—be delayed.

Questions about the Return, it ought to be realized, are not new; they have inexorably been linked in the past—as they are today—to the expectation and the apprehension of the Day of the

Lord, the End of Days, Armageddon. Four millennia ago, the Near East witnessed a god and his son promising Heaven on Earth. More than three millennia ago, king and people in Egypt yearned for a messianic time. Two millennia ago, the people of Judea wondered whether the Messiah had appeared, and we are still seized with the mysteries of those events. Are prophecies coming true?

We shall deal with the puzzling answers that were given, solve ancient enigmas, decipher the origin and meaning of symbols—the Cross, the Fishes, the Chalice. We shall describe the role of space-related sites in historic events, and show why Past, Present and Future converge in Jerusalem, the place of the "Bond Heaven-Earth." And we shall ponder why it is that our current twenty-first century A.D. is so similar to the twenty-first century B.C.E. Is history repeating itself—is it destined to repeat itself? Is it all guided by a Messianic Clock? Is the time at hand?

More than two millennia ago, Daniel of Old Testament fame repeatedly asked the angels: *When?* When will be the End of Days, the End of Time? More than three centuries ago the famed Sir Isaac Newton, who elucidated the secrets of celestial motions, composed treatises on the Old Testament's Book of Daniel and the New Testament's Book of Revelation; his recently found handwritten calculations concerning the End of Days will be analyzed, along with more recent predictions of The End.

Both the Hebrew Bible and the New Testament asserted that the secrets of the Future are embedded in the Past, that the destiny of Earth is connected to the Heavens, that the affairs and fate of Mankind are linked to those of God and gods. In dealing with what is yet to happen, we cross over from history to prophecy; one cannot be understood without the other, and we shall report them both. With that as our guide, let us look at what is to come through the lens of what had been. The answers will be certain to surprise.

ZECHARIA SITCHIN
New York, November 2006

Chapter I

THE MESSIANIC CLOCK

Wherever one turns, humankind appears seized with Apocalyptic trepidation, Messianic fervor, and End of Time anxiety.

Religious fanaticism manifests itself in wars, rebellions, and the slaughter of "infidels." Armies amassed by Kings of the West are warring with armies of the Kings of the East. A Clash of Civilizations shakes the foundations of traditional ways of life. Carnage engulfs cities and towns; the high and the mighty seek safety behind protective walls. Natural calamities and ever-intensifying catastrophies leave people wondering: Has Mankind sinned, is it witnessing Divine Wrath, is it due for another annihilating Deluge? Is this the Apocalypse? Can there be—will there be—Salvation? Are Messianic times afoot?

The time—the twenty-first century A.D.—or was it the twenty-first century B.C.E.?

The correct answer is Yes and Yes, both in our own time as well as in those ancient times. It is the condition of the present time, as well as at a time more than four millennia ago; and the amazing similarity is due to events in the middle time in between—the period associated with the messianic fervor at the time of Jesus.

Those three cataclysmic periods for Mankind and its planet—two in the recorded past (circa 2100 B.C.E. and when B.C.E. changed to A.D.), one in the nearing future—are interconnected; one has led to the other, one can be understood only by understanding the other. The Present stems from the Past, the Past is the Future. Essential to all three is **Messianic Expectation**; and linking all three is **Prophecy**.

How the present time of troubles and tribulations will end—what the Future portends—requires entering the realm of Prophecy. Ours will not be a mélange of newfound predictions whose main magnet is fear of doom and End, but a reliance upon unique ancient records that documented the Past, predicted the Future, and recorded previous Messianic expectations—prophesying the future in antiquity and, one believes, the Future that is to come.

In all three apocalyptic instances—the two that had occurred, the one that is about to happen—the physical and spiritual relationship between Heaven and Earth was and remains pivotal for the events. The physical aspects were expressed by the existence on Earth of actual sites that linked Earth with the heavens—sites that were deemed crucial, that were focuses of the events; the spiritual aspects have been expressed in what we call Religion. In all three instances, a changed relationship between Man and God was central, except that when, circa 2100 B.C.E., Mankind faced the first of these three epochal upheavals, the relationship was between men and *gods*, in the plural. Whether that relationship has really changed, the reader will soon discover.

The story of the gods, the **Anunnaki** ("Those who from heaven to Earth came"), as the Sumerians called them, begins with their coming to Earth from **Nibiru** in need of gold. The story of their planet was told in antiquity in the *Epic of Creation*, a long text on seven tablets; it is usually considered to be an allegorical myth, the product of primitive minds that spoke of planets as living gods combating each other. But as I have shown in my book *The Twelfth Planet*, the ancient text is in fact a sophisticated cosmogony that tells how a stray planet, passing by our solar system, collided with a planet called Tiamat; the collision resulted in the creation of Earth and its Moon, of the Asteroid Belt and comets, and in the capture of the invader itself in a great elliptical orbit that takes about 3,600 Earth-years to complete **(Fig. 1)**.

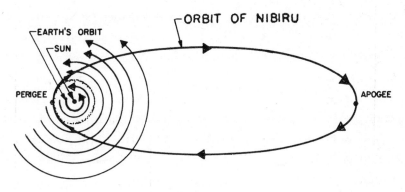

FIGURE I

It was, Sumerian texts tell, 120 such orbits—432,000 Earth-years—prior to the Deluge (the "Great flood") that the Anunnaki came to Earth. How and why they came, their first cities in the E.DIN (the biblical Eden), their fashioning of the Adam and the reasons for it, and the events of the catastrophic Deluge—have all been told in *The Earth Chronicles* series of my books, and will not be repeated here. But before we time-travel to the momentous twenty-first century B.C.E., some pre-Diluvial and post-Diluvial landmark events need to be recalled.

The biblical tale of the Deluge, starting in chapter 6 of Genesis, ascribes its conflicting aspects to a sole deity, Yahweh, who at first is determined to wipe Mankind off the face of the Earth, and then goes out of his way to save it through Noah and the Ark. The earlier Sumerian sources of the tale ascribe the disaffection with Mankind to the god **Enlil**, and the countereffort to save

THE END OF DAYS

Mankind to the god **Enki**. What the Bible glossed over for the sake of Monotheism was not just the disagreement between Enlil and Enki, but a rivalry and a conflict between two clans of Anunnaki that dominated the course of subsequent events on Earth.

That conflict between the two and their offspring, and the Earth regions allocated to them after the Deluge, need to be kept in mind to understand all that happened thereafter.

The two were half-brothers, sons of Nibiru's ruler **Anu**; their conflict on Earth had its roots on their home planet, Nibiru. Enki—then called **E.A** ("He whose home is water")—was Anu's firstborn son, but not by the official spouse, **Antu**. When Enlil was born to Anu by Antu—a half-sister of Anu—Enlil became the Legal Heir to Nibiru's throne though he was not the firstborn son. The unavoidable resentment on the part of Enki and his maternal family was exacerbated by the fact that Anu's accession to the throne was problematic to begin with: having lost out in a succession struggle to a rival named Alalu, he later usurped the throne in a coup d'état, forcing Alalu to flee Nibiru for his life. That not only backtracked Ea's resentments to the days of his forebears, but also brought about other challenges to the leadership of Enlil, as told in the epic *Tale of Anzu*. (For the tangled relationships of Nibiru's royal families and the ancestries of Anu and Antu, Enlil and Ea, see *The Lost Book of Enki*.)

The key to unlocking the mystery of the gods' succession (and marriage) rules was my realization that these rules also applied to the people chosen by them to serve as their proxies to Mankind. It was the biblical tale of the Patriarch Abraham explaining (*Genesis* 20:12) that he did not lie when he had presented his wife Sarah as his sister: "Indeed, she is my sister, the daughter of my father, but not the daughter of my mother, and she became my wife." Not only was marrying a half-sister from a different mother permitted, but a son by her—in this case Isaac—became the Legal Heir and dynastic successor, rather than the Firstborn Ishmael, the son of the handmaiden Hagar. (How such succession rules caused the bitter feud between Ra's divine descendants in Egypt, the

4

half-brothers Osiris and Seth who married the half-sisters Isis and Nephtys, is explained in *The Wars of Gods and Men.*)

Though those succession rules appear complex, they were based on what those who write about royal dynasties call "blood-lines"—what we now should recognize as sophisticated DNA genealogies that also distinguished between general DNA inherited from the parents as well as the mitochondrial DNA (mtDNA) that is inherited by females only from the mother. The complex yet basic rule was this: Dynastic lines continue through the male line; the Firstborn son is next in succession; a half-sister could be taken as wife *if she had a different mother*; and if a son by such a half-sister is later born, that son—though not Firstborn—becomes the Legal Heir and the dynastic successor.

The rivalry between the two half-brothers Ea/Enki and Enlil in matters of the throne was complicated by personal rivalry in matters of the heart. They both coveted their half-sister **Ninmah**, whose mother was yet another concubine of Anu. She was Ea's true love, but he was not permitted to marry her. Enlil then took over, and had a son by her—**Ninurta**. Though born without wedlock, the succesion rules made Ninurta Enlil's uncontested heir, being both his Firstborn son and one born by a royal half-sister.

Ea, as related in *The Earth Chronicles* books, was the leader of the first group of fifty Anunnaki to come to Earth to obtain the gold needed to protect Nibiru's dwindling atmosphere. When the initial plans failed, his half-brother Enlil was sent to Earth with more Anunnaki for an expanded Mission Earth. If that was not enough to create a hostile atmosphere, Ninmah too arrived on Earth to serve as chief medical officer . . .

A long text known as the *Atrahasis Epic* begins the story of gods and men on Earth with a visit by Anu to Earth to settle once and for all (he hoped) the rivalry between his two sons that was ruining the vital mission; he even offered to stay on Earth and let one of the half-brothers assume the regency on Nibiru. With that in mind, the ancient text tells us, lots were drawn to determine who would stay on Earth and who would sit on Nibiru's throne:

> The gods clasped hands together,
> had cast lots and had divided:
> Anu went up [back] to heaven,
> [For Enlil] the Earth was made subject;
> The seas, enclosed as with a loop,
> to Enki the prince were given.

The result of drawing lots, then, was that Anu returned to Nibiru as its king. Ea, given dominion over the seas and waters (in later times, "Poseidon" to the Greeks and "Neptune" to the Romans), was granted the epithet EN.KI ("Lord of Earth") to soothe his feelings; but it was EN.LIL ("Lord of the Command") who was put in overall charge: "To him the Earth was made subject." Resentful or not, Ea/Enki could not defy the rules of succession or the results of the drawing of lots; and so the resentment, the anger at justice denied, and a consuming determination to avenge injustices to his father and forefathers and thus to himself led Enki's son **Marduk** to take up the fight.

Several texts describe how the Anunnaki set up their settlements in the E.DIN (the post-Diluvial Sumer), each with a specific function, and all laid out in accordance with a master plan. The crucial space connection—the ability to constantly stay in communication with the home planet and with the shuttlecraft and spacecraft—was maintained from Enlil's command post in **Nippur**, the heart of which was a dimly lit chamber called the DUR. AN.KI, "The Bond Heaven-Earth." Another vital facility was a spaceport, located at Sippar ("Bird City"). Nippur lay at the center of concentric circles at which the other "cities of the gods" were located; all together they shaped out, for an arriving spacecraft, a landing corridor whose focal point was the Near East's most visible topographic feature—the twin peaks of Mount Ararat **(Fig. 2)**.

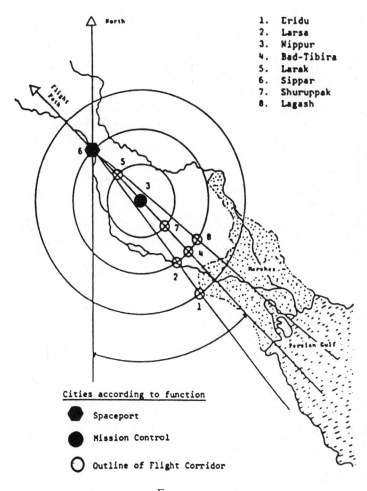

FIGURE 2

And then the Deluge "swept over the earth," obliterated all the cities of the gods with their Mission Control Center and Spaceport, and buried the Edin under millions of tons of mud and silt. Everything had to be done all over again—but much could no longer be the same. First and foremost, it was necessary to create a new spaceport facility, with a new Mission Control

Center and new Beacon-sites for a Landing Corridor. The new landing path was anchored again on the prominent twin peaks of Ararat; the other components were all new: the actual spaceport in the Sinai Peninsula, on the 30th parallel north; artificial twin peaks as beacon sites, the Giza pyramids; and a new Mission Control Center at a place called Jerusalem **(Fig. 3)**. It was a layout that played a crucial role in post-Diluvial events.

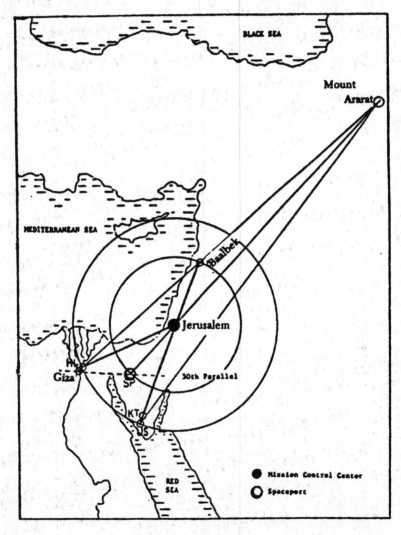

FIGURE 3

. . .

The Deluge was a watershed (both literally and figuratively) in the affairs of both gods and men, and in the relationship between the two: the Earthlings, who were fashioned to serve and work for the gods were henceforth treated as junior partners on a devastated planet.

The new relationship between men and gods was formulated, sanctified, and codified when Mankind was granted its first high civilization, in Mesopotamia, circa 3800 B.C.E. The momentous event followed a state visit to Earth by Anu, not just as Nibiru's ruler but also as the head of the pantheon, on Earth, of the ancient gods. Another (and probably the main) reason for his visit was the establishment and affirmation of peace among the gods themselves—a live-and-let-live arrangement dividing the lands of the Old World among the two principal Anunnaki clans, that of Enlil and that of Enki—for the new post-Diluvial circumstances and the new location of the space facilities required a new territorial division among the gods.

It was a division that was reflected in the biblical Table of Nations (*Genesis*, chapter 10), in which the spread of Mankind, emanating from the three sons of Noah, was recorded by nationality and geography: Asia to the nations/lands of Shem, Europe to the descendants of Japhet, Africa to the nation/lands of Ham. The historical records show that the parallel division among the gods allotted the first two to the Enlilites, the third one to Enki and his sons. The connecting Sinai peninsula, where the vital post-Diluvial spaceport was located, was set aside as a neutral Sacred Region.

While the Bible simply listed the lands and nations according to their Noahite division, the earlier Sumerian texts recorded the fact that the division was a deliberate act, the result of deliberations by the leadership of the Anunnaki. A text known as the *Epic of Etana* tells us that

The great Anunnaki who decree the fates
sat exchanging their counsels regarding the Earth.
They created the four regions,
set up the settlements.

In the First Region, the lands between the two rivers Euphrates and Tigris (Mesopotamia), Man's first known high civilization, that of Sumer, was established. Where the prediluvial cities of the gods had been, Cities of Man arose, each with its sacred precinct where a deity resided in his or her ziggurat—Enlil in Nippur, Ninmah in Shuruppak, Ninurta in Lagash, **Nannar/Sin** in Ur, **Inanna/Ishtar** in Uruk, **Utu/Shamash** in Sippar, and so on. In each such urban center an EN.SI, a "Righteous Shepherd"—initially a chosen demigod—was selected to govern the people in behalf of the gods; his main assignment was to promulgate codes of justice and morality. In the sacred precinct, a priesthood overseen by a high priest served the god and his spouse, supervised the holiday celebrations, and handled the rites of offerings, sacrifices, and prayers to the gods. Art and sculpture, music and dance, poetry and hymns, and above all writing and recordkeeping flourished in the temples and extended to the royal palace.

From time to time one of those cities was selected to serve as the land's capital; there the ruler was king, LU.GAL ("Great man"). Initially and for a long time thereafter this person, the most powerful man in the land, served as both king and high priest. He was carefully chosen, for his role and authority, and all the physical symbols of Kingship, were deemed to have come to Earth directly from Heaven, from Anu on Nibiru. A Sumerian text dealing with the subject stated that before the symbols of Kingship (tiara/crown and scepter) and of Righteousness (the shepherd's staff) were granted to an earthly king, they "lay deposited before Anu in heaven." Indeed, the Sumerian word for Kingship was *Anuship*.

This aspect of "Kingship" as the essence of civilization, just behavior and a moral code for Mankind, was explicitly expressed in the statement, in the *Sumerian King List*, that after the Deluge *"Kingship was brought down from Heaven."* It is a profound statement that must be borne in mind as we progress in this book to the messianic expectations—in the words of the New Testament, for the **Return of the "Kingship of Heaven" to Earth.**

Circa 3100 B.C.E. a similar yet not identical civilization was established in the Second Region in Africa, that of the river Nile (Nubia and Egypt). Its history was not as harmonious as that among the Enlilites, for rivalry and contention continued among Enki's six sons, to whom not cities but whole land domains were allocated. Paramount was an ongoing conflict between Enki's firstborn **Marduk** (*Ra* in Egypt) and **Ningishzidda** (*Thoth* in Egypt), a conflict that led to the exile of Thoth and a band of African followers to the New World (where he became known as *Quetzalcóatl*, the Winged Serpent). Marduk/Ra himself was punished and exiled when, opposing the marriage of his young brother Dumuzi to Enlil's granddaughter Inanna/Ishtar, he caused his brother's death. It was as compensation to Inanna/Ishtar that she was granted dominion over the Third Region of civilization, that of the Indus Valley, circa 2900 B.C.E. It was for good reason that the three civilizations—as was the spaceport in the sacred region—were all centered on the 30th parallel north **(Fig. 4)**.

According to Sumerian texts, the Anunnaki established Kingship—civilization and its institutions, as most clearly exemplified in Mesopotamia—as a new order in their relationships with Mankind, with kings/priests serving both as a link and a separator between gods and men. But as one looks back on that seemingly "golden age" in the affairs of gods and men, it becomes evident that the affairs of the gods constantly dominated and determined the affairs of Men and the fate of Mankind. Overshadowing all was the determination of Marduk/Ra to undo the injustice done to his father Ea/Enki, when under the succession rules of the Anunnaki not Enki but Enlil was declared the Legal Heir of their father Anu, the ruler on their home planet Nibiru.

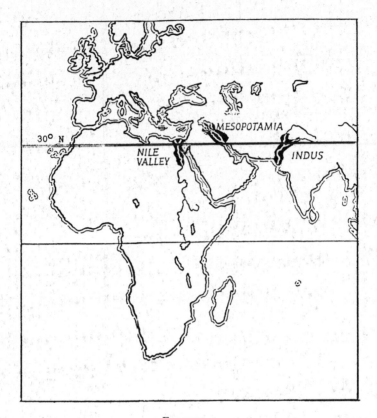

FIGURE 4

In accord with the sexagesimal ("base sixty") mathematical system that the gods granted the Sumerians, the twelve great gods of the Sumerian pantheon were given numerical ranks in which Anu held the supreme Rank of Sixty; the Rank of Fifty was granted to Enlil; that of Enki was forty, and so farther down, alternating between male and female deities **(Fig. 5)**. Under the succession rules, Enlil's son Ninurta was in line for the rank of fifty on Earth, while Marduk held a nominal rank of ten; and initially, these two successors-in-waiting were not yet part of the twelve "Olympians."

And so the long, bitter, and relentness struggle by Marduk that began with the Enlil–Enki feud focused later on Marduk's contention with Enlil's son Ninurta for the succession to the Rank of Fifty, and then extended to Enlil's granddaughter Inanna/

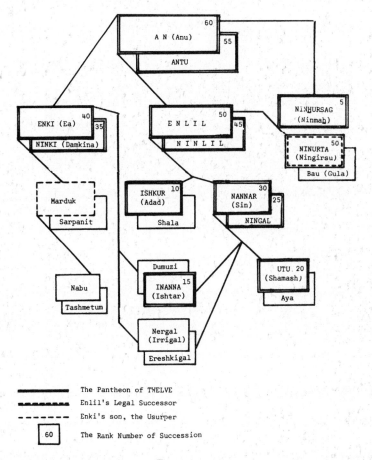

The Pantheon of TWELVE
Enlil's Legal Successor
Enki's son, the Usurper
The Rank Number of Succession

FIGURE 5

Ishtar, whose marriage to Dumuzi, Enki's youngest son, was so opposed by Marduk that it ended with Dumuzi's death. In time Marduk/Ra faced conflicts even with other brothers and half-brothers of his, in addition to the conflict with Thoth that we have already mentioned—principally with Enki's son **Nergal**, who married a granddaughter of Enlil named Ereshkigal.

In the course of these struggles, the conflicts at times flared up to full-fledged wars between the two divine clans; some of those wars are called "The Pyramid Wars" in my book *The Wars of Gods and Men*. In one notable instance the fighting led to the burying alive of Marduk inside the Great Pyramid; in another, it led to its capture by Ninurta. Marduk was also exiled more than once—both as punishment and as a self-imposed absence. His persistent efforts to attain the status to which he believed he was entitled included the event recorded in the Bible as the Tower of Babel incident; but in the end, after numerous frustrations, success came only when Earth and Heaven were aligned with the **Messianic Clock**.

Indeed, the first cataclysmic set of events, in the twenty-first century B.C.E., and the Messianic expectations that accompanied it, is principally the story of Marduk; it also brought to center stage his son **Nabu**—a deity, the son of a god, but whose mother was an Earthling.

Throughout the history of Sumer that spanned almost two thousand years, its royal capital shifted—from the first one, Kish (Ninurta's first city), to Uruk (the city that Anu granted to Inanna/Ishtar) to Ur (Sin's seat and center of worship); then to others and then back to the initial ones; and finally, for the third time, back to Ur. But at all times Enlil's city Nippur, his "cult center," as scholars are wont to call it, remained the religious center of Sumer and the Sumerian people; it was there that the annual cycle of worshipping the gods was determined.

The twelve "Olympians" of the Sumerian pantheon, each with his or her celestial counterpart among the twelve members of the Solar System (Sun, Moon, and ten planets, including Nibiru), were also honored with one month each in the annual cycle of a twelve-month year. The Sumerian term for "month," EZEN, actually meant holiday, festival; and each such month was devoted to celebrating the worship-festival of one of the twelve supreme gods. It was the need to determine the exact time when each such month began and ended (and not in order to enable peasants to know when to sow or harvest, as schoolbooks explain) that led to the introduction of *Mankind's first calendar* in **3760 B.C.E.** It is known as the **Calendar of Nippur** because it was the task of its priests to determine the calendar's intricate timetable and to announce, for the whole land, the time of the religious festivals. That calendar is still in use to this day as the Jewish religious calendar, which, in A.D. 2007, numbers the year as 5767.

In pre-Diluvial times Nippur served as Mission Control Center, Enlil's command post where he set up the DUR.AN.KI, the "Bond Heaven-Earth" for the communications with the home planet Nibiru and with the spacecraft connecting them. (After the Deluge, these functions were relocated to a place later known as Jerusalem.) Its central position, equidistant from the other functional centers in the E.DIN (see Fig. 2), was also deemed to be equidistant from the "four corners of the Earth" and gave it the nickname *"Navel of the Earth."* A hymn to Enlil referred to Nippur and its functions thus:

> *Enlil,*
> *When you marked off divine settlements on Earth,*
> *Nippur you set up as your very own city . . .*
> *You founded the Dur-An-Ki*
> *In the center of the four corners of the Earth.*

(The term "the Four Corners of the Earth" is also found in the Bible; and when Jerusalem replaced Nippur as Mission Control Center after the Deluge, it too was nicknamed the Navel of the Earth.)

In Sumerian the term for the four regions of the Earth was UB, but it also is found as AN.UB—the heavenly, the *celestial* four "corners"—in this case an astronomical term connected with the calendar. It is taken to refer to the four points in the Earth-Sun annual cycle that we nowadays call the Summer Solstice, the Winter Solstice, and the two crossings of the equator—once as the Spring Equinox and then as the Autumnal Equinox. In the Calendar of Nippur, the year began on the day of the Spring Equinox and it has so remained in the ensuing calendars of the ancient Near East. That determined the time of the most important festival of the year—the New Year festival, an event that lasted ten days, during which detailed and canonized rituals had to be followed.

Determining calendrical time by Heliacal Rising entailed the observation of the skies at dawn, when the sun just begins to rise on the eastern horizon but the skies are still dark enough to show the stars in the background. The day of the equinox having been determined by the fact that on it daylight and nighttime were precisely equal, the position of the sun at heliacal rising was then marked by the erection of a stone pillar to guide future observations—a procedure that was followed, for example, later on at Stonehenge in Britain; and, as at Stonehenge, long-term observations revealed that the group of stars ("constellation") in the background has not remained the same **(Fig. 6)**; there, the alignment stone called the "Heel Stone" that points to sunrise on solstice day nowadays, pointed originally to sunrise circa 2000 B.C.E.

The phenomenon, called Precession of the Equinoxes or just

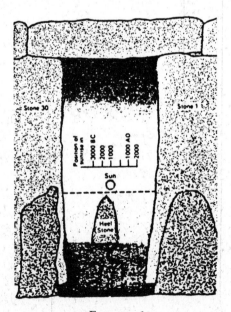

FIGURE 6

Precession, results from the fact that as the Earth completes one
annual orbit around the Sun, it does not return to the same exact
celestial spot. There is a slight, very slight retardation; it amounts
to one degree (out of 360 in the circle) in 72 years. It was Enki
who first grouped the stars observable from Earth into "constellations," and divided the heavens in which the Earth circled
the sun into twelve parts—what has since been called the Zodiacal Circle of constellations **(Fig. 7)**. Since each twelfth part of
the circle occupied 30 degrees of the celestial arc, the retardation
or Precessional shift from one Zodiacal House to another lasted

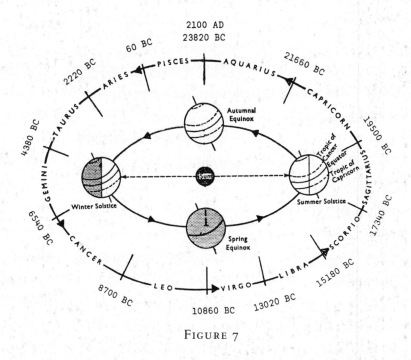

FIGURE 7

(mathematically) **2,160** years (72 × 30), and a complete zodiacal
cycle lasted 25,920 years (2,160 × 12). The approximate dates of the
Zodiacal Ages—following the equal twelve-part division and
not actual astronomical observations—have been added here for
the reader's guidance.

That this was the achievement from a time preceding Mankind's civilizations is attested to by the fact that a zodiacal calendar was applied to Enki's first stays on Earth (when the first two zodiacal houses were named in his honor); that this was not the achievement of a Greek astronomer (Hipparchus) in the third century B.C.E. (as most textbooks still suggest), is attested by the fact that the twelve zodiacal houses were known to the Sumerians millennia earlier by names **(Fig. 8)** and depictions **(Fig. 9)** that we use to this day.

1. GU.AN.NA ("heavenly bull"), *Taurus*.
2. MASH.TAB.BA ("twins"), our *Gemini*.
3. DUB ("pincers," "tongs"), the Crab or *Cancer*.
4. UR.GULA ("lion"), which we call *Leo*.
5. AB.SIN ("her father was Sin"), the Maiden, *Virgo*.
6. ZI.BA.AN.NA ("heavenly fate"), the scales of *Libra*.
7. GIR.TAB ("which claws and cuts"), *Scorpio*.
8. PA.BIL ("defender"), the Archer, *Sagittarius*.
9. SUHUR.MASH ("goat-fish"), *Capricorn*.
10. GU ("lord of the waters"), the Water Bearer, *Aquarius*.
11. SIM.MAH ("fishes"), *Pisces*.
12. KU.MAL ("field dweller"), the Ram, *Aries*.

FIGURE 8

In *When Time Began* the calendrical timetables of gods and men were discussed at length. Having come from Nibiru, whose orbital period, the SAR, meant 3,600 (Earth-) years, that unit was naturally the first calendrical yardstick of the Anunnaki even on the fast-orbiting Earth. Indeed, the texts dealing with their early days on Earth, such as the *Sumerian King Lists*, designated the periods of this or that leader's time on Earth in terms of Sars. I termed this **Divine Time**. The calendar granted to Mankind, one based on the orbital aspects of the Earth (and its Moon), was named **Earthly Time**. Pointing out that the 2,160-year zodiacal shift (less than a year for the Anunnaki) offered them a better

GIR.TAB
Scorpio

PA.BIL
Sagittarius

SUHUR.MASH
Capricorn

AB.SIN
Virgo .

FIGURE 9

ratio—the "golden ratio" of 10:6—between the two extremes; I called this **Celestial Time**.

As Marduk discovered, that Celestial Time was the "clock" by which his destiny was to be determined.

But which was **Mankind's Messianic Clock**, determining its fate and destiny—*Earthly Time*, such as the count of fifty-year Jubilees, a count in centuries, or the Millennium? Was it *Divine Time*, geared to Nibiru's orbit? Or was it—is it—*Celestial Time* that follows the slow rotation of the zodiacal clock?

The quandary, as we shall see, baffled mankind in antiquity; it still lies at the core of the current Return issue. The question that is posed has been asked before—by Babylonian and Assyrian stargazing priests, by biblical Prophets, in the Book of Daniel, in the Revelation of St. John the Divine, by the likes of Sir Isaac Newton, by all of us today.

The answer will be astounding. Let us embark on the painstaking quest.

Chapter II

"And It Came to Pass"

It is highly significant that in its record of Sumer and the early Sumerian civilization, the Bible chose to highlight the *space connection incident*—the one known as the tale of the "**Tower of Babel**":

> *And it came to pass as they journeyed from the east*
> *that they found a plain in the land of Shin'ar*
> *and they settled there.*
> *And they said to one another:*
> *"Come, let us make bricks and burn them by fire."*
> *And the brick served them as stone,*
> *and the bitumen served them as mortar.*
> *And they said: "Come, let us build us a city*
> *and* a tower whose head shall reach the heavens."
>
> GENESIS 11: 2–4

This is how the Bible recorded the most audacious attempt—by Marduk!—to assert his supremacy by establishing his own city in the heart of Enlilite domains and, moreover, to **build there his own space facility with its own *launch tower*.**

The place is named in the Bible *Babel,* "Babylon" in English.

This biblical tale is remarkble in many ways. It records, first of all, the settlement of the Tigris-Euphrates plain after the Deluge, after the soil had dried up enough to permit resettlement. It correctly names the new land *Shin'ar,* the Hebrew name for Sumer. It provides the important clue from where—from the mountainous region to the east—the settlers had come. It recognizes that it was there that Man's first urban civilization began—the building of cities. It correctly notes (and explains) that in that land, where the soil consisted of layers of dried mud and there is no native rock, the people used mud bricks for building and by hardening the bricks in kilns could use them instead of stone. It also refers to the use of bitumen as mortar in construction—an astounding bit of information, since bitumen, a natural petroleum product, seeped up from the ground in southern Mesopotamia but was totally absent in the Land of Israel.

The authors of this chapter in Genesis were thus well informed regarding the origins and key innovations of the Sumerian civilization; they also recognized the significance of the "Tower of Babel" incident. As in the tales of the creation of Adam and of the Deluge, they melded the various Sumerian deities into the plural *Elohim* or into an all-encompassing and supreme *Yahweh,* but they left in the tale the fact that it took a *group of deities* to say, "let **us** come down" and put an end to this rogue effort (*Genesis* 11:7).

Sumerian and later Babylonian records attest to the veracity of the biblical tale and contain many more details, linking the incident to the overall strained relationships between the gods that caused the outbreak of two "Pyramid Wars" after the Deluge. The "Peace on Earth" arrangements, circa 8650 B.C.E., left the erstwhile *Edin* in Enlilite hands. That conformed to the decisions of Anu, Enlil, and even Enki—but was never acquiesced to by Marduk/Ra. And so it was that when Cities of Men began to be allocated in the former *Edin* to the gods, Marduk raised the issue, "What about me?"

Although Sumer was the heartland of the Enlilite territories and its cities were Enlilite "cult centers," there was one exception: in the south of Sumer, at the edge of the marshlands, there was **Eridu**; it was rebuilt after the Deluge at the exact same site where Ea/Enki's first settlement on Earth had been. It was Anu's insistence, when the Earth was divided among the rival Anunnaki clans, that Enki forever retain Eridu as his own. Circa **3460 B.C.E.** Marduk decided that he could extend his father's privilege to also having his own foothold in the Enlilite heartland.

The available texts do not provide the reason why Marduk chose that specific site on the banks of the Euphrates river for his new headquarters, but its location provides a clue: it was situated between the rebuilt Nippur (the pre-Diluvial Mission Control Center) and the rebuilt Sippar (the pre-Diluvial spaceport of the Anunnaki), so what Marduk had in mind could have been a facility that served both functions. A later map of Babylon, drawn on a clay tablet **(Fig. 10)** represents it as a "Navel of the Earth"—akin to Nippur's original function-title. The name Marduk gave the place, *Bab-Ili* in Akkadian, meant "Gateway of the gods"—a place from which the gods could ascend and descend, where the appropriate main facility was to be a "tower whose head shall reach the heavens"—a **launch tower**!

As in the biblical tale, so it is told in parallel (and earlier) Mesopotamian versions that this attempt to establish a rogue space facility came to naught. Though fragmented, the Mesopotamian texts (first translated by George Smith in 1876) make it clear that Marduk's act infuriated Enlil, who "in his anger a command poured out" for a nighttime attack to destroy the tower.

Egyptian records report that a chaotic period that lasted 350 years preceded the start of Pharaonic kingship in Egypt, circa **3110 B.C.E.** It is this time frame that leads us to date the Tower of Babel incident to circa **3460 B.C.E.**, for the end of that chaotic period marked the return of Marduk/Ra to Egypt, the expulsion of Thoth, and the start of the worship of Ra.

Figure 10

Frustrated this time, Marduk never gave up his attempts to dominate the official space facilities that served as the "Bond Heaven-Earth," the link between Nibiru and Earth—or to set up his own facility. Since, in the end, Marduk did attain his aims in Babylon, the interesting question is: Why did he fail in **3460 B.C.E.**? The equally interesting answer is: It was a matter of timing.

A well-known text recorded a conversation between Marduk and his father, Enki, in which a disheartened Marduk asked his father what he had failed to learn. What he failed to do was to take into account the fact that the time then—the Celestial Time—was **the Age of the Bull, the Age of Enlil.**

. . .

Among the thousands of inscribed tablets unearthed in the ancient Near East, quite a number provided information regarding
the month associated with a particular deity. In a complex calendar begun in Nippur in **3760 B.C.E.**, the first month, *Nissanu,*
was the EZEN (festival time) for Anu and Enlil (in a leap year
with a thirteenth lunar month, the honor was split between the
two). The list of "honorees" changed as time went by, as did the
composition of the membership of the supreme Pantheon of
Twelve. The month associations also changed locally, not only in
various lands but sometimes to recognize the city god. We know,
for example, that the planet we call Venus was initially associated
with Ninmah and later on with Inanna/Ishtar.

Though such changes make difficult the identifications of who
was linked celestially to what, some zodiacal associations can be
clearly inferred from texts or drawings. Enki (called at first E.A,
"He whose home is water") was clearly associated with the Water
Bearer "Aquarius" **(Fig. 11)**, and initially if not permanently also
with the Fishes, "Pisces." The constellation that was named The
Twins, "Gemini," without doubt was so named in honor of the

FIGURE 11

24

only known divine twins born on Earth—Nannar/Sin's children Utu/Shamash and Inanna/Ishtar. The feminine constellation of "Virgo" (the "Maiden" rather than the inaccurate "Virgin") that, like the planet Venus, was probably named at first in honor of Ninmah, was renamed AB.SIN, "Whose father is Sin," which could be correct only for Inanna/Ishtar. The Archer or Defender, "Sagittarius," matched the numerous texts and hymns extolling Ninurta as the Divine Archer, his father's warrior and defender. Sippar, the city of Utu/Shamash, no longer the site of a spaceport after the Deluge, was considered in Sumerian times to be the center of Law and Justice, and the god was deemed (even by the later Babylonians) as the Chief Justice of the land; it is certain that the Scales of Justice, "Libra," represented his constellation.

And then there were the nicknames comparing the prowess, strength, or characteristics of a god with an animal held in awe; Enlil's, as text after text reiterated, was the *Bull*. It was depicted on cylinder seals, on tablets dealing with astronomy, and in art. Some of the most beautiful art objects discovered in the Royal Tombs of Ur were bull heads sculpted in bronze, silver, and gold, adorned with semiprecious stones. Without doubt, the constellation of the Bull—Taurus—honored and symbolized Enlil. Its name, GUD.ANNA, meant "The Bull of Heaven," and texts dealing with an actual "Bull of Heaven" linked Enlil and his constellation to one of the most unique places on Earth.

It was a place that was called *The Landing Place*—and it is there that one of the most amazing structures on Earth, including a stone tower that reaches to the heavens, still stands.

Many texts from antiquity, including the Hebrew Bible, describe or refer to the unique forest of tall and great cedar trees in Lebanon. In ancient times it extended for miles, surrounding the unique place—a *vast stone platform built by the gods as their first space-related site on Earth*, before their centers and real spaceport were established. It was, Sumerian texts attested, the only structure that had survived the Deluge, and could thus serve right after the Deluge as a base of operations for the Anunnaki; from it they revived the ravished lands with crops and domesticated animals. The place, called the "Landing Place" in the *Epic of Gilgamesh*, was that king's destination in his search for immortality; we learn from the epic tale that it was there, in the sacred cedar forest, that Enlil kept the GUD.ANNA—the "Bull of Heaven," the symbol of Enlil's Age of the Bull.

And what happened then in the sacred forest had a bearing on the course of the affairs of gods and men.

The journey to the Cedar Forest and its Landing Place, we learn from the epic tale, began in Uruk, the city that Anu granted as a present to his great-granddaughter Inanna (a name that meant "Beloved of Anu"). Its king, early in the third millennium B.C.E., was **Gilgamesh (Fig. 12)**. He was no ordinary man, for his mother was the goddess Ninsun, a member of Enlil's family. That made Gilgamesh not a mere *demi*-god, but one who was "*two thirds* divine." As he got older and began to contemplate matters of life and death, it occurred to him that being two-thirds divine

FIGURE 12

26

ought to make a difference; why should he "peer over the wall" like an ordinary mortal? he asked his mother. She agreed with him, but explained to him that the apparent immortality of the gods was in reality longevity due to the long orbital period of their planet. To attain such longevity he had to join the gods on Nibiru; and to do that, he had to go to the place where the rocket ships ascend and descend.

Though warned of the journey's hazards, Gilgamesh was determined to go. If I fail, he said, at least I will be remembered as one who had tried. At his mother's insistence an artificial double, Enkidu (ENKI.DU meant "By Enki Made"), was to be his companion and guardian. Their adventures, told and retold in the Epic's twelve tablets and its many ancient renderings, can be followed in our book *The Stairway to Heaven.* There were, in fact, not one but two journeys **(Fig. 13)**: one was to the Landing Place in the Cedar Forest, the other to the spaceport in the Sinai peninsula where—according to Egyptian depictions **(Fig. 14)**—rocket ships were emplaced in underground silos.

In the first journey circa **2860 B.C.E.**—to the Cedar Forest in Lebanon—the duo were assisted by the god Shamash, the godfather of Gilgamesh, and the going was relatively quick and easy. After they reached the forest *they witnessed during the night the launching of a rocket ship.* This is how Gilgamesh described it:

> *The vision that I saw was wholly awesome!*
> *The heavens shrieked, the earth boomed.*
> *Though daylight was dawning, darkness came.*
> *Lightning flashed, a flame shot up.*
> *The clouds swelled, it rained death!*
> *Then the glow vanished, the fire went out,*
> *And all that had fallen was turned to ashes.*

Awed but undeterred, the next day Gilgamesh and Enkidu discovered the secret entrance that had been used by the Anunnaki,

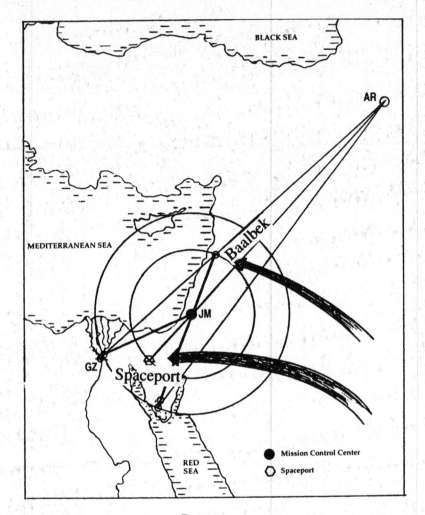

FIGURE 13

but as soon as they entered it, they were attacked by a robotlike guardian who was armed with death beams and a revolving fire. They managed to destroy the monster, and relaxed by a brook thinking that their way in was clear. But when they ventured deeper into the Cedar Forest, a new challenger appeared: **the Bull of Heaven.**

Unfortunately, the sixth tablet of the epic is too damaged for the lines describing the creature and the battle with it to be completely legible. The legible portions do make it clear that the two

FIGURE 14

comrades ran for their lives, pursued by the Bull of Heaven all the way back to Uruk; it was there that Enkidu managed to slay it. The text becomes legible where the boastful Gilgamesh, who cut off the bull's thigh, "called the craftsmen, the armorers, the artisans" of Uruk to admire the bull's horns. The text suggests that they were *artificially made*—"each is cast from thirty minas of lapis, the coating on each is two fingers thick."

Until another tablet with the illegible lines is discovered, we shall not know for sure whether Enlil's celestial symbol in the cedar forest was a specially selected living bull decorated and embellished with gold and precious stones or a robotic creature, an artificial monster. What we do know for certain is that upon its slaying, "Ishtar, in her abode, set up a wail" all the way to Anu in the heavens. The matter was so serious that Anu, Enlil, Enki, and Shamash formed a divine council to judge the comrades (only Enkidu ended up being punished) and to consider the slaying's consequences.

The ambitious Inanna/Ishtar had indeed reason to raise a wail: the invincibility of Enlil's Age had been pierced, and the Age itself was symbolically shortened by the cutting off of the bull's thigh. We know from Egyptian sources, including pictorial depictions in astronomical papyri **(Fig. 15)**, that the slaying's

FIGURE 15

symbolism was not lost on Marduk: it was taken to mean that in the heavens, too, the Age of Enlil had been cut short.

Marduk's attempt to establish an alternative space facility was not taken lightly by the Enlilites; the evidence suggests that Enlil and Ninurta were preoccupied with establishing their own alternative space facility on the other side of the Earth, in the Americas, near the post-Diluvial sources of gold.

This absence, together with the Bull of Heaven incident, ushered in a period of instability and confusion in their Mesopotamian heartland, subjecting it to incursions from neighboring lands. People called Gutians, then the Elamites came from the East; Semitic-speaking peoples came from the West. But while the Easterners worshiped the same Enlilite gods as the Sumerians, the *Amurru* ("Westerners") were different. Along the shores of the "Upper Sea" (the Mediterranean), in the lands of the Canaanites, the people were beholden to the Enki'ite gods of Egypt.

Therein lay the seeds—perhaps to this day—of Holy Wars undertaken *"In the Name of God,"* except that different peoples had different national gods . . .

It was Inanna who came up with a brilliant idea; it can be described as "if you can't fight them, invite them in." One day, as she was roaming the skies in her Sky Chamber—it happened circa **2360 B.C.E.**—she landed in a garden next to a sleeping man who had caught her fancy. She liked the sex, she liked the man. He was a Westerner, speaking a Semitic language. As he wrote later in his memoirs, he knew not who his father was, but knew that his mother was an *Entu*, a god's priestess, who put him in a reed basket that was carried by the river's flowing waters to a garden tended by Akki the Irrigator, who raised him as a son.

The possibility that the strong and handsome man could have been a god's castoff son was enough for Inanna to recommend to the other gods that the next king of the land should be this Amurru. When they agreed, she granted him the epithet-name *Sharru-kin*, the old cherished title of Sumerian kings. Not stemming from the previous recognized royal Sumerian lineages, he could not ascend the throne in any one of the olden capitals, and a brand-new city was established to serve as his capital. It was called *Aggade*—"Union City." Our textbooks call this king Sargon of Akkad and his Semitic language Akkadian. His kingdom, which added northern and northwestern provinces to ancient Sumer, was called *Sumer & Akkad*.

Sargon lost little time in carrying out the mission for which he was selected—to bring the "rebel lands" under control. Hymns to Inanna—henceforth known by the Akkadian name *Ishtar*—had her tell Sargon that he would be remembered "by the destruction of the rebel land, massacring its people, making its rivers run with blood." Sargon's military expeditions were recorded and glorified in his own royal annals; his achievements were summarized in the *Sargon Chronicle* thus:

Sharru-kin, king of Aggade,
Rose to power in the era of Ishtar.
He left neither rival nor opponent.
He spread his terror-inspiring awe in all the lands.
He crossed the sea in the east,
He conquered the country of the west
in its full extent.

The boast implies that the sacred space-related site, the Landing Place deep in the "country of the west," was captured and held in behalf of Inanna/Ishtar—but not without opposition. Even texts written in glorification of Sargon state that "in his old age all the provinces revolted against him." Counterannals, recording the events as viewed from Marduk's side, reveal that Marduk led a punishing counteroffensive:

On account of the sacrilege Sargon committed,
the great god Marduk became enraged . . .
From east to west he alienated the people from Sargon,
and punished him with an affliction of being without rest.

Sargon's territorial reach, it needs to be noted, included only one of the four post-Diluvial space-related sites—only the Landing Place in the Cedar Forest (see Fig. 3). Sargon was briefly succeeded on the throne of Sumer & Akkad by two sons, but his true successor in spirit and deed was a grandson named Naram-Sin. The name meant "Sin's favorite," but the annals and inscriptions concerning his reign and military campaigns show that he was in fact Ishtar's favorite. Texts and depictions record that Ishtar encouraged the king to seek grandeur and greatness by ceaseless conquest and destruction of her enemies, actively assisting him on the battlefields. Depictions of her, which used to show her as an enticing goddess of love, now showed her as a goddess of war, bristling with weapons **(Fig. 16)**.

FIGURE 16

It was warfare not without a plan—a plan to counter Marduk's ambitions by capturing *all* the space-related sites in behalf of Inanna/Ishtar. The lists of cities captured or subdued by Naram-Sin indicate that he not only reached the Mediterranean Sea—assuring control of the Landing Place—but also turned southward to invade Egypt. Such an incursion into the Enki'ite domains was unprecedented, and it could take place, a careful reading of the records reveals, because Inanna/Ishtar had formed an unholy alliance with Nergal, Marduk's brother who espoused Inanna's sister. The thrust into Egypt also required entering and crossing the neutral Sacred Region in the Sinai Peninsula, where the spaceport was located—another breach of the olden Peace Treaty. Boastful, Naram-Sin gave himself the title "King of the four regions" . . .

We can hear the protests of Enki. We can read texts that record Marduk's warnings. It was all more than even the Enlilite leadership could condone. A long text known as *The Curse of Aggade*, which tells the story of the Akkadian dynasty, clearly states that its end came about "after the frowning of the forehead of Enlil." And so the "word of Ekur"—the decision of Enlil from his temple in Nippur—was to put an end to it: "The word of the Ekur was upon Aggade" to be destroyed and wiped off the face of the Earth. Naram-Sin's end came circa **2260 B.C.E.**; texts from that time report that troops from the territory in the east, called Gutium, loyal to Ninurta, were the instrument of divine wrath; Aggade was never rebuilt, never resettled; that royal city, indeed, has never been found.

The saga of Gilgamesh at the start of the third millennium B.C.E., and the military forays of the Akkadian kings near the end of that millennium, provide a clear background for that millennium's events: the targets were the space-related sites—by Gilgamesh to attain the gods' longevity, by the kings beholden to Ishtar to attain supremacy.

Without doubt, it was Marduk's "Tower of Babel" attempt that placed the control of the space-related sites at the center of the affairs of gods and men; and as we shall see, that centrality dominated much (if not most) of what took later place.

The Akkadian phase of the War and Peace on Earth was not without celestial or "messianic" aspects.

In his chronicles, Sargon's titles followed the customary honorific "Overseer of Ishtar, king of Kish, great Ensi of Enlil," but he also called himself "**anointed priest of Anu.**" It was the first time that being divinely *anointed*—which is what "Messiah" literally means—appears in ancient inscriptions.

Marduk, in his pronouncements, warned of coming upheavals and cosmic phenomena:

> *The day shall be turned into darkness,*
> *the flow of river waters shall be disarrayed,*
> *the lands shall be laid to waste,*
> *the people will be made to perish.*

Looking back, recalling similar biblical prophecies, it is clear that on the eve of the twenty-first century B.C.E., gods and men expected a coming Apocalyptic Time.

Chapter III

EGYPTIAN PROPHECIES, HUMAN DESTINIES

In the annals of Man on Earth, the twenty-first century B.C.E. saw in the ancient Near East one of civilization's most glorious chapters, known as the Ur III period. It was at the same time the most difficult and crushing one, for it witnessed the end of Sumer in a deathly nuclear cloud. And after that, nothing was the same.

Those momentous events, as we shall see, were also the root of the messianic manifestations that centered on Jerusalem when B.C.E. turned to A.D. some twenty-one centuries later.

The historic events of that memorable century—as all events in history—had their roots in what had taken place before. Of that, the year **2160 B.C.E.** is a date worth remembering. The annals of Sumer & Akkad from that time record a major policy shift by the Enlilite gods. In Egypt, the date marked the beginning of changes of political-religious significance, and what occurred in both zones coincided with a new phase in Marduk's campaign to attain supremacy. Indeed, it was Marduk's chesslike strategy maneuvers and geographic movements from one place to another that controlled the agenda of the era's "divine chess game." His moves and movements began with a departure from Egypt, to become (in Egyptian eyes) **Amon** (also written *Amun* or *Amen*), ***"The Unseen."***

The date of **2160 B.C.E.** is considered by Egyptologists to mark the beginning of what is designated the First Intermediate Period—a chaotic interval between the end of the Old Kingdom and the dynastic start of the Middle Kingdom. During the thousand years of the Old Kingdom, when the religious-political capital was Memphis in Middle Egypt, the Egyptians worshipped the Ptah pantheon, erecting monumental temples to him, to his son Ra, and to their divine successors. The famed inscriptions of the Memphite Pharaohs glorified the gods and promised an Afterlife for the kings. Reigning as the gods' surrogates, those Pharaohs wore the double crown of Upper (southern) and Lower (northern) Egypt, signifying not just the administrative but also the religious unification of the Two Lands, unification attained when Horus defeated Seth in their struggle for the Ptah/Ra legacy. And then, in **2160 B.C.E.**, that unity and religious certainty came crashing down.

The turmoil saw a breakup of the Union, abandonment of the capital, attacks from the south by Theban princes to gain control, foreign incursions, desecration of temples, a collapse of law and order, and droughts, famines, and food riots. Those conditions are recalled in a papyrus known as the *Admonitions of Ipu-Wer,* a long hieroglyphic text that consists of several sections in which it gives an account of calamities and tribulations, blames an unholy enemy for religious wrongdoing and social evils, and calls on the people to repent and resume the religious rites. A prophetic section describing the *coming of a Redeemer,* and another that extolls the ideal times that will follow, conclude the papyrus.

At its start the text describes the breakdown of law and order and of a functioning society—a situation in which "the door-keepers go and plunder, the wash-man refuses to carry his load . . . robbery is everywhere . . . a man regards his son as an enemy." Though the Nile is in flooding and irrigates the land, "no one

ploughs . . . grain has perished . . . the storehouses are bare . . . dust
is throughout the land . . . the desert spreads . . . women are dried
up, no one can conceive . . . the dead are just thrown into the
river . . . the river is blood." The roads are unsafe, trade has
ceased, the provinces of Upper Egypt are no longer taxed; "there
is civil war . . . barbarians from elsewhere have come to Egypt . . .
all is in ruin."

Some Egyptologists believe that at the core of those events lay
a simple rivalry for wealth and power, an attempt (successful in
the end) by Theban princes from the south to control and rule the
whole country. Lately, studies have associated the collapse of the
Old Kingdom with a "climate change" that undermined a society
founded on agriculture, caused food shortages and food riots, social upheaval, and collapse of authority. But little attention has
been paid to a major and perhaps the most important change: in
the texts, in the hymns, in the honorific names of temples, it was
no longer Ra but from then on Ra-*Amon*, or simply *Amon*, who
was henceforth worshipped; Ra became *Amon*—Ra the Unseen—
for he was gone from Egypt.

It was indeed a religious change that caused the political and
societal breakdown, the unidentified Ipu-Wer wrote; we believe
that the change was Ra's becoming Amon. The upheaval began
with a collapse of religious observances and manifested itself in
the defiling and abandonment of temples, where "the Place of
Secrets has been laid bare, the writings of the august enclosure
have been scattered, common men tear them up in the streets . . .
magic is exposed, it is in the sight of him who knows it not." The
sacred symbol of the gods worn on the king's crown, the Uraeus
(the Divine Serpent), "is rebelled against . . . religious dates are
disturbed . . . priests are carried off wrongfully."

After calling on the people to repent, "to offer incense in the
temples . . . to keep the offerings to the gods," the papyrus called
on the repenters to *be baptized*—to "remember to immerse." Then
the words of the papyrus turn prophetic: in a passage that even

Egyptologists call "truly messianic," the Admonitions speak of "a time that shall come" when an unnamed *Savior—a "god-king"—* shall appear. Starting with a small following, of him "men shall say:

> *He brings coolness upon the heart,*
> *He is a shepherd of all men.*
> *Though his herds may be small,*
> *He will spend the days caring for them . . .*
> *Then he would smite down evil,*
> *He would stretch forth his arm against it."*

"People will be asking: 'Where is he today? Is he then sleeping? Why is his power not seen?'" Ipu-Wer wrote, and answered, "Behold, the glory thereof cannot be seen, [but] Authority, Perception and Justice are with him."

Those ideal times, Ipu-Wer stated in his prophecy, will be preceded by their own messianic birth pangs: "Confusion will set throughout the land, in tumultuous noise one will kill the other, the many will kill the few." People will ask: "Does the Shepherd desire death?" No, he answered, "it is the land that commands death," but after years of strife, righteousness and proper worship will prevail. This, the papyrus concluded, was "What Ipu-Wer said when he responded to the majesty of the All-Lord."

If not just the description of events and the messianic prophecies, but also the choice of wording in that ancient Egyptian papyrus seem astounding, there is more to come. Scholars are aware of the existence of another prophetic/messianic text that reached us from ancient Egypt, but believe that it was really composed after the events and only pretends to be prophetic by dating itself to an earlier time. To be specific, while the text purports to relate prophecies made at the time of Sneferu, a Fourth Dynasty pharaoh (circa 2600 B.C.E.), Egyptologists believe that it was actually written in the time of Amenemhet I of the Twelfth Dynasty (circa 2000 B.C.E.)—*after* the events that it pretends to prophecy.

Even so, the "prophecies" serve to confirm those prior occurrences; and many details and the very wording of the predictions can best be described as chilling.

The prophecies are purported to be told to King Sneferu by a "great seer-priest" named Nefer-Rohu, "a man of rank, a scribe competent with his fingers." Summoned to the king to foretell the future, Nefer-Rohu "stretched forth his hand for the box of writing equipment, he drew forth a scroll of papyrus," and then began to write what he was envisioning, in a Nostradamus-like manner:

> *Behold, there is something about which men speak;*
> *It is terrifying . . .*
> *What will be done was never done before.*
> *The Earth is completely perished.*
> *The land is damaged, no remainder exists.*
> *There is no sunshine that people could see,*
> *No one can live with the covering clouds,*
> *The south wind opposes the north wind.*
> *The rivers of Egypt are empty . . .*
> *Ra must begin the foundations of the Earth again.*

Before Ra can restore the "Foundations of the Earth," there will be invasions, wars, bloodshed. Then a new era of peace, tranquility, and justice will follow. It will be brought by what we have come to call a Savior, a Messiah:

> *Then it is that a sovereign will come—*
> *Ameni ("The Unknown"),*
> *The Triumphant he will be called.*
> *The Son-Man will be his name forever and ever . . .*
> *Wrongdoing will be driven out;*
> *Justice in its place will come;*
> *The people of his time rejoice.*

**It is astounding to find such messianic prophecies of apoc-
alyptic times and the end of Wrongdoing that will be followed by the coming—the return—of peace and justice, in
papyrus texts written some 4,200 years ago; it is chilling to
find in them terminology that is familiar from the New
Testament, about an Unknown, the Triumphant Savior,
the "Son-Man."**

It is, as we shall see, a link in millennia-spanning interconnected events.

In Sumer, a period of chaos, occupation by foreign troops, defiling of temples, and confusion as to where the capital should be
and who should be king followed the end of the Sargonic Era of
Ishtar in **2260 B.C.E.**

For a while, the only safe haven in the land was Ninurta's
"cult center" Lagash, from which the Gutian foreign troops were
kept out. Mindful of Marduk's unrelenting ambitions, Ninurta
decided to reassert his right to the Rank of Fifty by instructing
the then-king of Lagash, Gudea, to erect for him in the city's
Girsu (the sacred precinct) a new and different temple. Ninurta—
here called NIN.GIRSU, "Lord of the Girsu"—already had a
temple there, as well as a special enclosure for his "Divine Black
Bird" or flying machine. Yet the building of the new temple required special permission from Enlil, which was in time granted.
We learn from the inscriptions that the new temple had to have
special features linking it to the heavens, enabling certain celestial observations. To that end Ninurta invited to Sumer the god
Ningishzidda ("Thoth" in Egypt), the Divine Architect, and
Keeper of the Secrets of the Giza pyramids. The fact that Ning-
ishzidda/Thoth was the brother whom Marduk forced into exile
circa 3100 B.C.E. was certainly not lost on all concerned . . .

The amazing circumstances surrounding the announcement,
planning, construction, and dedication of the E.NINNU ("House/

Temple of Fifty") are told in great detail in Gudea's inscriptions; they were unearthed in the ruins of Lagash (a site now called Tello) and are quoted at length in *The Earth Chronicles* books. What emerges from that detailed record (inscribed on two clay cylinders in clear Sumerian cuneiform script, **Fig. 17**) is the fact that from announcement to dedication, every step and every detail of the new temple was dictated by celestial aspects.

Those special celestial aspects had to do with the very timing of the temple's building: It was the time, as the inscriptions' open-

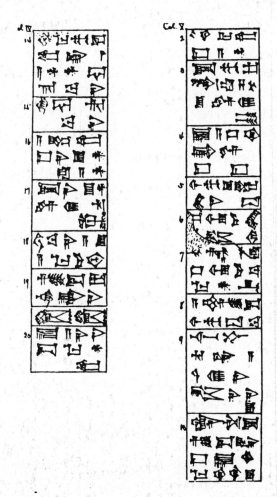

FIGURE 17

ing lines declare, when "in the heavens destinies on Earth were determined":

> At the time when in heaven
> destinies on Earth were determined,
> "Lagash shall lift its head heavenwards
> in accordance with the Great Tablet of Destinies"
> Enlil in favor of Ninurta decided.

That special time when the destinies on Earth are determined in the heavens was what we have called Celestial Time, the Zodiacal Clock. That such determining was linked to Equinox Day becomes evident from the rest of Gudea's tale, as well as from Thoth's Egyptian name *Tehuti, The Balancer* (of day and night) who "Draws the Cord" for orienting a new temple. Such celestial considerations continued to dominate the Eninnu project from start to finish.

Gudea's tale begins with a vision-dream that reads like an episode from *The Twilight Zone* TV series, for while the several gods featured in it were gone when he awoke, the various objects they showed him in the dream remained physically lying by his side!

In that vision-dream (the first of several) the god Ninurta appeared at sunrise, and the sun was aligned with the planet Jupiter. The god spoke and informed Gudea that he was chosen to build a new temple. Next the goddess Nisaba appeared; she was wearing the image of a temple structure on her head; the goddess was holding a tablet on which the starry heavens were depicted, and with a stylus she kept pointing to the "favorable celestial constellation." A third god, Ningishzidda (i.e. Thoth) held a tablet of lapis lazuli on which a structural plan was drawn; he also held a clay brick, a mold for brickmaking, and a builder's carrying basket. When Gudea awoke, the three gods were gone, but the architectural tablet was on his lap **(Fig. 18)** and the brick and its mold were at his feet!

Gudea needed the help of an oracle goddess and two more vision-dreams to understand the meaning of it all. In the third

FIGURE 18

vision-dream he was shown a holographic-like animated demonstration of the temple's building, starting with the initial alignment with the indicated celestial point, the laying of foundations, the molding of bricks—the construction all the way up, step by step. Both the start of construction and the final dedication ceremony were to be held on signals from the gods on specific days; both fell on New Year's Day, which meant the day of the Spring Equinox.

The temple "raised its head" in the customary seven stages, but—unusually for the flat-topped Sumerian ziggurats—its head had to be pointed, "shaped like a horn"—Gudea had to emplace upon the temple's top a capstone! Its shape is not described, but in all probabilty (and judging by the image on Nisaba's head), it was in the shape of a pyramidion—in the manner of capstones on Egyptian pyramids **(Fig. 19)**. Moreover, rather than leave the brickwork exposed, as was customary, Gudea was required to encase the structure with a casing of reddish stones, increasing its similarity to an Egyptian pyramid. "The outside view of the temple was like that of a mountain set in place."

That raising a structure with the appearance of an Egyptian

FIGURE 19

pyramid had a purpose becomes clear from Ninurta's own words. The new temple, he told Gudea, "will be seen from afar; its awe-inspiring glance will reach the heavens; the adoration of my temple shall extend to all the lands, its heavenly name will be proclaimed in countries from the ends of the Earth—

In Magan and Meluhha it will cause people [to say]:
Ningirsu [the "Lord of the Girsu"],
the Great Hero from the Lands of Enlil,
is a god who has no equal;
He is the lord of all the Earth.

Magan and Meluhha were the Sumerian names for Egypt and Nubia, the Two Lands of the gods of Egypt. The purpose of the Eninnu was to establish, even there, in Marduk's lands, Ninurta's unequaled Lordship: "A god who has no equal, the Lord of all the Earth."

Proclaiming Ninurta's (rather than Marduk's) supremacy required special features in the Eninnu. The ziggurat's entrance had to face the Sun precisely in the east, rather than the customary northeast. In the temple's topmost level Gudea had to erect a SHU. GA.LAM—"where the shining is announced, the place of the aperture, the place of determining," from which Ninurta/Ningirsu could see "the Repetition over the lands." It was a circular chamber with twelve positions, each marked with a zodiacal symbol, with an aperture for observing the skies—*an ancient planetarium aligned to the zodiacal constellations!*

In the temple's forecourt, linked to an avenue that faced sunrise, Gudea had to erect two stone circles, one with six and the other with seven stone pillars, for observing the skies. Since only one avenue is mentioned, one assumes that the circles were one within the other. As one studies each phrase, terminology, and structural detail, it becomes evident that what was built in Lagash with the help of Ningishzidda/Thoth was a complex yet practical stone observatory, one part of which, devoted entirely to the zodiacs, reminds one of the similar one found in Denderah, Egypt **(Fig. 20)**, and the

FIGURE 20

other, geared to observing celestial risings and setting, **a virtual Stonehenge on the banks of the Euphrates river!**

Like Stonehenge in the British Isles **(Fig. 21)**, the one built in Lagash provided stone markers for solar observations of solstices and equinoxes, but the prime outside feature was the creation of a sight line from a center stone, continued between two stone pillars, then on down an avenue to another stone. Such a sight line, precisely oriented when planned, enabled determining at the moment of heliacal rising in which zodiacal constellation the Sun was appearing. *And that—determining the zodiacal age through precise observation—was the prime objective of the whole complex facility.*

In Stonehenge, that sight line ran (and still runs) from the stone column called the Altar Stone in the center, through two stone columns identified as Sarsen Stones numbers 1 and 30, then down the Avenue to the so-called Heel Stone (see Fig. 6). It is generally agreed that the Stonehenge with the double Bluestone Circle and

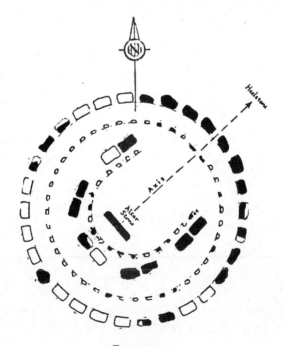

FIGURE 21

46

the Heel Stone of what is designated Stonehenge II dates to between 2200 to 2100 B.C.E. *That was also the time—perhaps more accurately, in 2160 B.C.E.—when the "Stonehenge on the Euphrates" was built.*

And that was no chance coincidence. Like those two zodiacal observatories, other stone observatories proliferated at the same time in other places on Earth—at various sites in Europe, in South America, on the Golan Heights northeast of Israel, even in faraway China (where archaeologists discovered in the Shanzi province a stone circle with thirteen pillars aligned to the zodiac and dating to 2100 B.C.E.). They were all deliberate countermoves by Ninurta and Ningishzidda to Marduk's Divine Chess Game: **to show Mankind that the zodiacal age was still the Age of the Bull.**

Various texts from that time, including an autobiographical text by Marduk and a longer text known as the *Erra Epos,* shed light on Marduk's wanderings away from Egypt, making him there the Hidden One. They also reveal that his demands and actions assumed an urgency and ferocity because of a conviction that his time for supremacy has come. The Heavens bespeak *my* glory as Lord, was his claim. Why? Because, he announced, the Age of the Bull, the Age of Enlil, was over; *the Age of the Ram, Marduk's zodiacal age, has arrived.* It was, just as Ninurta had told Gudea, the time when in the heavens destinies on Earth were determined.

The zodiacal ages, it will be recalled, were caused by the phenomenon of Precession, the retardation in Earth's orbit around the Sun. The retardation accumulates to 1 degree (out of 360) in 72 years; an arbitrary division of the grand circle into 12 segments of 30 degrees each means that mathematically the zodiacal calendar shifts from one Age to another every 2,160 years. Since the Deluge occurred, according to Sumerian texts, in the Age of the Lion, our zodiacal clock can start circa **10860 B.C.E.**

An astounding timetable emerges if, in this *mathematically de-termined* 2,160-year zodiacal calendar, the starting point of 10800 B.C. rather than 10860 B.C. is chosen:

> 10800 to 8640—Age of the Lion (Leo)
> 8640 to 6480—Age of the Crab (Cancer)
> 6480 to 4320—Age of the Twins (Gemini)
> 4320 to 2160—Age of the Bull (Taurus)
> 2160 to 0—Age of the Ram (Aries)

Setting aside the neat end result that *synchronizes with the Christian Era*, one must wonder whether it was mere coincidence that the Ishtar-Ninurta era petered out in or about 2160 B.C.E., just when, according to the above zodiacal calendar, the Age of the Bull, Enlil's Age, was also ending? Probably not; certainly Marduk did not think so. The available evidence suggests that he was sure that according to Celestial Time, *his* time for supremacy, his Age, has arrived. (Modern studies of Mesopotamian astronomy indeed confirm that the zodiacal circle was divided there into 12 houses of 30 degrees each—a mathematical rather than an observational division.)

The various texts we have mentioned indicate that as he moved about, Marduk made another foray into the Enlilite heartland, arriving back in Babylon with a retinue of followers. Rather than resort to armed conflict, the Enlilites enlisted Marduk's brother Nergal (whose spouse was a granddaughter of Enlil) to come to Babylon from southern Africa and persuade his brother to leave. In his memoirs, known as *The Erra Epos,* Nergal reported that Marduk's chief argument was that his time, the Age of the Ram, had arrived. But Nergal counterargued that it is not really so: the Heliacal Rising, he told Marduk, still occurs in the constellation of the Bull!

Enraged, Marduk questioned the accuracy of the observations. What happened to the precise and reliable instruments, from before the Deluge, that were installed in your Lower World domain? he demanded to know from Nergal. Nergal explained that they were destroyed by the Deluge. Come, see for yourself which constellation is seen at sunrise on the appointed day, he urged Marduk. Whether Marduk went to Lagash to make the observation, we do not know, but he did realize the cause of the discrepancy:

While mathematically the ages changed every 2,160 years, in reality, observationally, they did not. The zodiacal constellations, in which stars were grouped arbitrarily, were not of equal size. Some occupied a larger arc of the heavens, some smaller; and as it happened, the constellation of the Ram was one of the smaller ones, squeezed between the larger Taurus and Pisces **(Fig. 22)**. Celestially, the constellation Taurus, occupying more than 30 degrees of the heavenly arc, lingers on for at least another two centuries beyond its mathematical length.

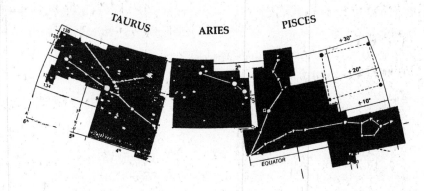

FIGURE 22

In the twenty-first century B.C.E., Celestial Time and Messianic Time failed to coincide.

Go away peacefully and come back when the heavens will declare your Age, Nergal told Marduk. Yielding to his fate, Marduk did leave, but did not go too far away.

And with him, as emissary, spokesman, and herald, was his son, whose mother was an Earthling woman.

Chapter IV

OF GODS AND DEMIGODS

The decision of Marduk to stay in or near the contested lands and to involve his son in the struggle for Mankind's allegiance persuaded the Enlilites to return Sumer's central capital to Ur, the cult center of Nannar (Su-en or *Sin* in Akkadian). It was the third time that Ur was chosen to serve in that capacity—hence the designation "Ur III" for that period.

The move linked the affairs of the contending gods to the biblical tale—and role—of Abraham, and the intertwined relationship changed Religion to this day.

Among the many reasons for the choice of Nannar/Sin as the Enlilite champion was the realization that contending with Marduk has expanded beyond the affairs of the gods alone, and has become a contest for the minds and hearts of the people—of the very Earthlings whom the gods had created, who now made up the armies that went to war on behalf of their creators . . .

Unlike other Enlilites, Nannar/Sin was not a combatant in the Wars of the Gods; his selection was meant to signal to people everywhere, even in the "rebel lands," that under his leadership an era of peace and prosperity would begin. He and his spouse **Ningal (Fig. 23)** were greatly beloved by the people of Sumer,

FIGURE 23

and Ur itself spelled prosperity and well-being; its very name, which meant "urban, domesticated place," came to mean not just "city" but The City—the urban jewel of the ancient lands.

Nannar/Sin's temple there, a skyscraping ziggurat, rose in stages within a walled sacred precinct where a variety of structures served as the gods' abode and the residences and functional buildings of a legion of priests, officials, and servants who attended to the divine couple's needs and arranged the religious observances by king and people. Beyond those walls there extended a magnificent city with two harbors and canals linking it to the Euphrates river **(Fig. 24)**, a great city with the king's palace, administrative buildings (including for scribes and record-keeping as well as for tax collecting), multilevel private dwellings, workshops, schools, merchants' warehouses, and stalls—all in wide streets where, at many intersections, prayer shrines open to all travelers were built. The majestic ziggurat with its monumental stairways (Reconstruction, **Fig. 25**), though long in ruins, still dominates the landscape even after more than 4,000 years.

But there was another compelling reason. Unlike the contending

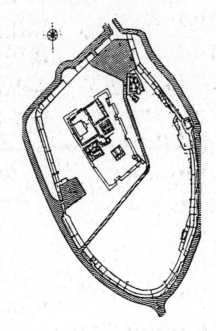

FIGURE 24

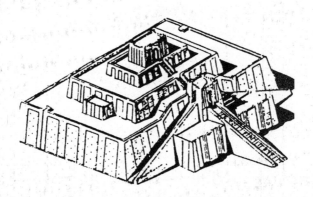

FIGURE 25

Ninurta and Marduk, who were both "immigrants" to Earth from Nibiru, Nannar/Sin was born on Earth. He was not only Enlil's Firstborn on Earth—he was the first of the first generation of gods to be born on Earth. His children, the twins Utu/Shamash and Inanna/Ishtar, and their sister Ereshkigal, who belonged to the gods' third generation, were all born on Earth. They were gods, but they were also Earth's natives. All that was without doubt taken into consideration in the coming struggle for the loyalties of the people.

The choice of a new king, to restart afresh kingship in and from Sumer, was also carefully made. Gone was the free hand given to (or assumed by) Inanna/Ishtar, who chose Sargon the Akkadian to start a new dynasty because she liked his lovemaking. The new king, named Ur-Nammu ("The joy of Ur"), was carefully selected by Enlil and approved by Anu, and he was no mere Earthling: He was a son—"the beloved son"—of the goddess Ninsun; she had been, the reader will recall, the mother of Gilgamesh. Since this divine genealogy was restated in numerous inscriptions during Ur-Nammu's reign, in the presence of Nannar and other gods, one must assume that the claim was factual. This made Ur-Nammu not only a demigod but—as was the case of Gilgamesh—"two-thirds divine." Indeed, the claim that the king's mother was the goddess Ninsun placed Ur-Nammu in the very same status as that of Gilgamesh, whose exploits were well remembered and whose name remained revered. The choice was thus a signal, to friends and foes alike, that the glorious days under the unchallenged authority of Enlil and his clan are back.

All that was important, perhaps even crucial, because Marduk had his own attributes of appeal to the masses of Mankind. That special appeal to the Earthlings was the fact that Marduk's deputy and chief campaigner was his son **Nabu**—who not only was born on Earth, but *was born to a mother who herself was an Earthling,* for long ago—indeed, in the days before the Deluge—Marduk broke all traditions and taboos and took an Earthling woman to be his official wife.

That young Anunnaki took Earthling females as wives should not come as a shocking surprise, for it is recorded in the Bible for all to read. What is little known even to scholars, because the information is found in ignored texts and has to be verified from complex God Lists, is the fact that it was Marduk who set the example that the "Sons of the gods" followed:

> *And it came to pass*
> *when the Earthlings began to increase in number*
> *upon the Earth*
> *and daughters were born unto them—*
> *That the sons of the* Elohim
> *saw the daughters of The Adam*
> *that they were compatible;*
> *And they took unto themselves wives*
> *of whichever they chose.*
>
> GENESIS 6: 1–2

The biblical explanation of the reasons for the Great Flood in the first eight enigmatic verses of chapter 6 of *Genesis* clearly points to the intermarriage and its resulting offspring as the cause of the divine wrath:

> *The* Nefilim *were on the Earth*
> *in those days and thereafter too,*
> *When the sons of the* Elohim
> *came unto the daughters of The Adam*
> *and had children by them.*

(My readers may recall that it was my question, as a schoolboy, of why *Nefilim*—which literally means "Those who have come down," who descended [from heaven to Earth]—was usually translated "giants." It was much later that I realized and suggested that the Hebrew word for "giants," *Anakim*, was actually a rendering of the Sumerian *Anunnaki*.)

The Bible clearly cites such intermarriage—the *"taking as wives"*— between young "sons of the gods" (sons of the *Elohim*, the *Nefilim*) and female Earthlings ("daughters of *The Adam*") as God's reason for seeking Mankind's end by the Deluge: "My spirit shall no longer dwell in Man, for in his flesh they erred . . . And God repented that He had fashioned the Adam on Earth, and was distraught, and He said: Let me wipe the Adam that I have created off the face of the Earth."

The Sumerian and Akkadian texts telling the story of the Deluge explained that two gods were involved in that drama: it was Enlil who sought Mankind's destruction by the Deluge, while it was Enki who connived to prevent it by instructing "Noah" to build the salvaging ark. When we delve into the details, we find that Enlil's "I've had it up to here!" anger on one hand, and Enki's counterefforts on the other hand, were not just a matter of principles. *For it was Enki himself who began to copulate with female Earthlings and have children by them*, and it was Marduk, Enki's son, who led the way to and set the example for actual marriages with them . . .

By the time their Mission Earth was fully operative, the Anunnaki stationed on Earth numbered 600; in addition, 300 who were known as the IGI.GI ("Those who observe and see") manned a planetary Way Station—on Mars!—and the spacecraft shuttling between the two planets. We know that Ninmah, the Anunnaki's chief medical officer, came to Earth at the head of a group of female nurses **(Fig. 26)**. It is not stated how many they were or

FIGURE 26

whether there were other females among the Anunnaki, but it is clear that in any event females were few among them. The situation required strict sexual rules and supervision by the elders, so much so that (according to one text) Enki and Ninmah had to act as matchmakers, decreeing who should marry whom.

Enlil, a strict disciplinarian, himself fell victim to the shortage of females and date-raped a young nurse. For that even he, the Commander in Chief on Earth, was punished with exile; the punishment was commuted when he agreed to marry Sud and make her his official consort, **Ninlil**. She remained his sole spouse to the very end.

Enki, on the other hand, is described in numerous texts as a philanderer with female goddesses of all ages, and managing to get away with it. Moreover, once "daughters of The Adam" proliferated, he was not averse to having sexual flings with them, too . . . Sumerian texts extolled Adapa, "the wisest of men" who grew up at Enki's household, was taught writing and mathematics by Enki, and was the first Earthling to be taken aloft to visit Anu on Nibiru; the texts also reveal that Adapa was a secret son of Enki, mothered by an Earthling female.

Apocryphal texts inform us that when Noah, the biblical hero of the Deluge, was born, much about the baby and the birth caused his father, Lamech, to wonder whether the real father had not been one of the Nefilim. The Bible just states that Noah was a genealogically "perfect" man who "Walked with the Elohim"; Sumerian texts, where the Flood's hero is named Ziusudra, suggest that he was a demigod son of Enki.

It was thus that one day Marduk complained to his mother that while his companions were assigned wives, he was not: "I have no wife, I have no children." And he went on to tell her that he had taken a liking to the daughter of a "high priest, an accomplished musician" (there is reason to believe that he was the chosen man Enmeduranki of Sumerian texts, the parallel of the biblical Enoch). Verifying that the young Earthling female—her name was Tsarpanit—agreed, Marduk's parents gave him the go-ahead.

The marriage produced a son. He was named EN.SAG, "Lofty Lord." But unlike Adapa, who was an Earthling demigod, Marduk's son was included in the Sumerian God Lists, where he was also called "the divine MESH"—a term used (as in GilgaMESH) to denote a demigod. *He was thus the first demigod who was a god.* Later on, when he led the masses of humans in his father's behalf, he was given the epithet-name **Nabu**—The Spokesman, *The Prophet*—for that is what the literal meaning of the word is, as is the meaning of the parallel biblical Hebrew word *Nabih,* translated "prophet."

Nabu was thus the god-son and an Adam-son of ancient scriptures, the one whose very name meant Prophet. As in the Egyptian prophecies earlier quoted, his name and role became linked to the Messianic expectations.

And so it was, in the days before the Deluge, that Marduk set an example to the other young unespoused gods: find and marry an Earthling female . . . The breach of the taboo appealed in particular to the Igigi gods who were away on Mars most of the time, with their principal station on Earth being the Landing Place in the Cedar Mountains. Finding an opportunity—perhaps an invitation to come and celebrate Marduk's wedding—they seized Earthling females and carried them off as wives.

Several extra-biblical books, designated The Apocrypha, such as the *Book of Jubilees,* the *Book of Enoch,* and the *Book of Noah,* record the incident of the intermarriage by the Nefilim and fill in the details. Some two hundred "Watchers" ("Those who observe and see") organized themselves in twenty groups; each group had a named leader. One, called Shamyaza, was in overall command. The instigator of the transgression, "the one who led astray the sons of God and *brought them down to Earth* and led them astray through the Daughters of Man," was named Yeqon . . . It happened, these sources confirmed, during the time of Enoch.

In spite of their efforts to fit the Sumerian sources (that told of rival and contradicting Enlil and Enki) into a monotheistic framework—the belief in only one Almighty God—the compilers of the Hebrew Bible ended that section in chapter 6 of Genesis with a recognition of the factual outcome. Speaking of the offspring of those intermarriages, the Bible makes two admissions: the first, that the intermarrying took place in the days before the Deluge, *"and therafter too"*; and secondly, that from the offspring "came the *heroes of old, the men of renown.*" The Sumerian texts indicate that post-Diluvial heroic kings were indeed such demigods.

But they were the offspring not only of Enki and his clan: sometimes kings in the Enlilite region were sons of Enlilite gods. For example, *The Sumerian King Lists* clearly state that when kingship began in Uruk (an Enlilite domain), the one chosen for kingship was a MESH, a demigod:

> *Meskiaggasher, a son of Utu,*
> *became high priest and king.*

Utu was of course the god Utu/Shamash, grandson of Enlil. Further down the dynastic line there was the famed Gilgamesh, "two thirds of him divine," son of the Enlilite goddess Ninsun and fathered by the High Priest of Uruk, an Earthling. (There were several more rulers down the line, both in Uruk and in Ur, who bore the title "Mesh" or "Mes".)

In Egypt, too, some Pharaohs claimed divine parentage. Many in the 18th and 19th Dynasties adopted theophoric names with a prefix or suffix MSS (rendered Mes, Mose, Meses), meaning "Issue of" this or that god—such as the names *Ah-mes* or *Ra-mses* (RA-MeSeS—"issue of," offspring of, the god Ra). The famed queen Hatshepsut, who though a female seized the title and privileges of a Pharaoh, claimed that right by virtue of being a demigod—the great god Amon, she claimed in inscriptions and depictions in her immense temple at Deir el Bahri, "took the form of his majesty the king," the husband of her queen-mother, "had intercourse with her," and caused Hatshepsut to be born as his semidivine daughter. Canaanite texts included the tale of Keret, a king who was the son of the god El.

An interesting variant on such demigod-as-king practices was the case of Eannatum, a Sumerian king in Ninurta's Lagash during the early "heroic" times. An inscription by the king on a well-known monument of his (the "Stela of the Vultures") attributes his demigod status to *artificial insemination* by Ninurta (the Lord of the Girsu, the sacred precinct), and to help from Inanna/Ishtar and Ninmah (here called by her epithet Ninharsag):

> *The Lord Ningirsu, warrior of Enlil,*
> *implanted the semen of Enlil for Eannatum*
> *in the womb of [. . .].*
> *Inanna accompanied his [birth],*
> *named him "Worthy in the Eanna temple,"*
> *set him on the sacred lap of Ninharsag..*
> *Ninharsag offered him her sacred breast.*
> *Ningirsu rejoiced over Eannatum—*
> *semen implanted in the womb by Ningirsu.*

While the reference to the "semen of Enlil" leaves unclear whether Ninurta/Ningirsu's own semen is here considered "semen of Enlil" because he was Enlil's firstborn, or actually used Enlil's semen for the insemination (which is doubtful), the inscription clearly claims that Eannatum's mother (whose name is illegible on the stela) was artificially impregnated, so that a demigod was conceived without actual sexual intercourse—**a case of immaculate conception in third millennium** B.C.E. **Sumer!**

That the gods were no strangers to artificial insemination is corroborated by Egyptian texts, according to which after Seth

killed and dismembered Osiris, the god Thoth extracted semen
from the phallus of Osiris and impregnated with it the wife of
Osiris, Isis, bringing about the birth of the god Horus. A depiction of the feat shows Thoth and birth goddesses holding the two
strands of DNA that were used, and Isis holding the newborn
Horus **(Fig. 27)**.

FIGURE 27

Clearly, then, after the Deluge the Enlilites too accepted both
the mating with Earthling females and considered the offspring
"heroes, men of renown," suitable for kingship.

Royal "bloodlines" of demigods were thus begun.

One of the first tasks of Ur-Nammu was to carry out a moral and
religious revival. And for that, too, a former revered and remembered king was emulated. It was done through the promulgation
of a new Code of Laws, laws of moral behavior, laws of justice—
of adherence, the Code said, to the laws that Enlil and Nannar

and Shamash had wanted the king to enforce and the people to live by.

The nature of the laws, a list of do's and don'ts, can be judged by Ur-Nammu's claim that due to those laws of justice, "the orphan did not fall prey to the wealthy, the widow did not fall prey to the powerful, the man with one sheep was not delivered to the man with one ox . . . justice was established in the land." In that he emulated—sometimes using the exact same phrases—a previous Sumerian king, Urukagina of Lagash, who three hundred years earlier had promulgated a law code by which social, legal, and religious reforms were instituted (among them the establishment of women's safehouses under the patronage of the goddess Bau, Ninurta's spouse). These, it ought to be pointed out, were the very same principles of justice and morality that the biblical prophets demanded of kings and people in the next millenium.

As the era of Ur III began, there was obviously a deliberate attempt to return Sumer (now Sumer & Akkad) to its olden days of glory, prosperity, and morality and peace—the times that preceded the latest confrontation with Marduk.

The inscriptions, the monuments, and the archaeological evidence attest that Ur-Nammu's reign, which began in 2113 B.C.E., witnessed extensive public works, restoration of river navigation, and the rebuilding and protection of the country's highways: "He made the highways run from the lower lands to the upper lands," an inscription stated. Greater trade and commerce followed. There was a surge in arts, crafts, schools, and other improvements in social and economic life (including the introduction of more accurate weights and measures). Treaties with neighboring rulers to the east and northeast spread the prosperity and well-being. The great gods, especially Enlil and Ninlil, were honored with renovated and magnified temples, and for the first time in Sumer's history, the priesthood of Ur was combined with that of Nippur, leading a religious revival.

All scholars agree that in virtually every way the Ur III period begun by Ur-Nammu attained new heights in the Sumerian civilization. That conclusion only increased the puzzlement caused by a beautifully crafted box that was uncovered by archaeologists: its inlaid panels, front and back, depicted two contradicting scenes of life in Ur. While one of the panels (now known as the "Peace Panel") depicted banqueting, commerce, and other scenes of civil activities, the other (the "War Panel") depicted a military column of armed and helmeted soldiers and horse-drawn chariots marching to war **(Fig. 28)**.

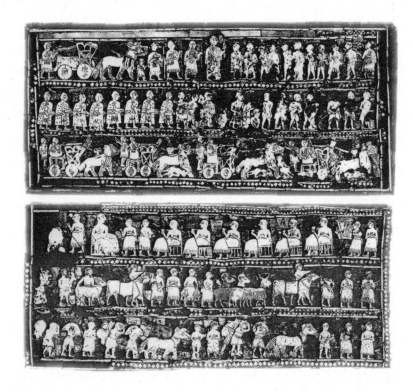

FIGURE 28

A close examination of the records from that time reveals that indeed while under the leadership of Ur-Nammu Sumer itself flourished, the hostility to the Enlilites by the "rebel lands" increased rather than diminished. The situation apparently demanded action, for accordng to Ur-Nammu's inscriptions Enlil gave him a "divine weapon that heaps up the rebels in piles" with which to attack "the hostile lands, destroy the evil cities and clear them of opposition." Those "rebel lands" and "sinning cities" were west of Sumer, the lands of Marduk's Amorite followers; there, the "evil"— the hostility against Enlil—was fanned by Nabu, who moved about from city to city proselytizing for Marduk. Enlilite records called him "The Oppressor," of whose influence the "sinning cities" had to be rid.

There is reason to believe that the Peace and War panels actually depicted Ur-Nammu himself—one showing him banqueting and celebrating peace and prosperity, the other in the royal chariot, leading his army to war. His military expeditions took him well beyond Sumer's borders into the western lands. But Ur-Nammu—great reformer, builder, and economic "shepherd" that he was—failed as a military leader. In the midst of battle his chariot got stuck in the mud; Ur-Nammu fell off it, but "the chariot like a storm rushed along," leaving the king behind, "abandoned like a crushed jug." The tragedy was compounded when the boat returning Ur-Nammu's body to Sumer "in an unknown place had sunk; the waves sank it down, with him on board."

When news of the defeat and the tragic death of Ur-Nammu reached Ur, a great lament went up there. The people could not understand how such a religiously devout king, a righteous shepherd who only followed the gods' directives with weapons they put in his hands, could perish so ignominiously. "Why did the Lord Nannar not hold him by the hand?" they asked; "Why did Inanna, Lady of Heaven, not put her noble arm around his head? Why did the valiant Utu not assist him?"

The Sumerians, who believed that all that happens had been fated, wondered, "Why did these gods step aside when Ur-Nammu's bitter fate was decided?" Surely those gods, Nannar and his twin children, knew what Anu and Enlil were determining; yet they said nothing to protect Ur-Nammu. There could be only one plausible explanation, the people of Ur and Sumer concluded as they cried out and lamented: The great gods must have gone back on their word—

> *How the fate of the hero had been changed!*
> *Anu altered his holy word.*
> *Enlil deceitfully changed his decree!*

These are strong words, accusing the great Enlilite gods of deceit and double-crossing! The ancient words convey the extent of the people's disappointment.

If that was so in Sumer & Akkad, one can imagine the reaction in the rebellious western lands.

In the struggle for the hearts and minds of Mankind, the Enlilites were faltering. Nabu, the "spokesman," intensified the campaign in behalf of his father Marduk. His own status was enhanced and changed: his own divinity was now glorified by a variety of venerating epithets. Inspired by Nabu—the *Nabih*, the Prophet—prophecies of the Future, of what is about to happen, began to sweep the contested lands.

We know what they said because a number of clay tablets on which such prophecies were inscribed have been found; written in Old Babylonian cuneiform, they are grouped by scholars as *Akkadian Prophecies* or *Akkadian Apocalypses*. **Common to all of them is the view that Past, Present, and Future are parts of a continuous flow of events;** that within a preordained Destiny there is some room for free will and thus a variated Fate; that for Mankind, both were decreed or determined by the gods of Heaven and Earth; and that *therefore events on Earth reflect occurrences in the heavens.*

To grant the prophecies believability, the texts sometimes anchored the prediction of future events in a known past historic occurrence or entity. What is wrong in the present, why change is needed, is then recounted. The unfolding events are attributed to decisions by one or more of the great gods. *A divine Emissary, a Herald, will appear*; the prophetic text might be his words, written down by the scribe, or expected pronouncements; as often as not, "a son will speak for his father." The predicted event(s) will be linked to omens—the death of a king, or heavenly signs: a celestial body will appear and make a frightful sound; "a burning fire" will come from the skies; "a star shall flash from the height of the sky to the horizon as a torch;" and, **most significantly, "a planet will appear before its time."**

Bad things, Apocalypse, shall precede the final event. There would be calamitous rains, huge devastating waves—or droughts, the silting of canals, locusts, and famines. Mother will turn against daughter, neighbor against neighbor. Rebellion, chaos, and calamities will occur in the lands. Cities will be attacked and depopulated; kings will die, be toppled, and captured; "one throne will overthrow another." Officials and priests will be killed; temples will be abandoned; rites and offerings will cease. And then the predicted event—a great change, a new era, a new leader, a Redeemer—will come. Good will prevail over evil, prosperity will replace sufferings; abandoned cities will be resettled, the remnants of the dispersed people will return to their homes. Temples will be restored, and the people will perform the correct religious rites.

Not unexpectedly, these Babylonian or pro-Marduk prophecies pointed the accusing finger of wrongdoing at Sumer & Akkad (and also their allies Elam, Hattiland, and the Sealands), and named the Amurru Westerners as the instrument of divine retribution. The Enlilite "cult centers" Nippur, Ur, Uruk, Larsa, Lagash, Sippar, and Adab are named; they will be attacked, plundered, their temples abandoned. The Enlilite gods are described as confused ("unable to sleep"). Enlil is calling out to Anu, but ignores Anu's advice (some translators read the word as "command") that Enlil issue a *misharu* edict—a "putting things straight" order. Enlil, Ishtar, and Adad will be forced to change kingship in Sumer & Akkad. The "sacred rites" will be transferred out of Nippur. Celestially, "the great planet" will appear in the constellation of the Ram. The word of Marduk shall prevail; "He will subdue the Four Regions, the whole Earth shall tremble at the mention of his name . . . After him his son will reign as king and will become master of the whole Earth."

In some of the prophecies, certain deities are the subject of specific predictions: "A king will arise," one text prophesied in regard to Inanna/Ishtar, "he will remove the protective goddess of Uruk from Uruk and make her dwell in Babylon . . . He will establish the rites of Anu in Uruk." The Igigi are also specifically mentioned: "The regular offerings for the Igigi gods, which had ceased, will be reestablished," one prophecy states.

As was the case with Egyptian prophecies, most scholars also treat the "Akkadian Prophecies" as "pseudo-prophecies" or *post aventum* texts—that they were in fact written long after the "predicted" events; but as we have remarked in regard to the Egyptian texts, to say that the events were not prophesied because they had already happened is only to reassert that the events per se did happen (whether or not they were predicted), and that is what matters most to us. *It means that the prophecies did come true.*

And if so, most chilling is the prediction (in a text known as *Prophecy "B"*):

> *The Awesome Weapon of Erra*
> *upon the lands and the people*
> *will come in judgment.*

A most chilling prophecy indeed, for before the twenty-first century B.C.E. was over, "judgment upon lands and peoples" occurred when the god Erra ("The Annihilator")—an epithet for Nergal—unleashed nuclear weapons in a cataclysm that made prophecies come true.

Chapter V

COUNTDOWN TO DOOMSDAY

The disastrous twenty-first century B.C.E. began with the tragic and untimely death of Ur-Nammu, in 2096 B.C.E. It culminated with an unparalleled calamity, by the hand of the gods themselves, in 2024 B.C.E. The interval was seventy-two years—exactly the precessional shift of one degree; and if it was just a coincidence, then it was one of a series of "coincidental" occurrences that were somehow well coordinated . . .

Following Ur-Nammu's tragic death, the throne of Ur was taken over by his son Shulgi. Unable to claim the status of a demigod, he asserted (in his inscriptions) that he was nevertheless born under divine auspices: the god Nannar himself arranged for the child to be conceived in Enlil's temple in Nippur through a union between Ur-Nammu and Enlil's high priestess, so that "a 'little Enlil,' a child suitable for kingship and throne, shall be conceived."

That was a genealogical claim not to be sneezed at. Ur-Nammu himself, as earlier stated, was "two-thirds" divine, since his mother was a goddess. Though the High Priestess who was Shulgi's mother is not named, her very status suggests that she, too, was of some godly lineage, for it was a king's daughter who was chosen to be an

EN.TU; and the kings of Ur, starting with the first dynasty, could be traced back to demigods. That Nannar himself arranged for the union to take place in Enlil's temple in Nippur was also significant; as previously stated, it was under Ur-Nammu's reign that for the first time the priesthood of Nippur was combined with the priesthood of another city—in this case, with the one in Ur.

Much of what was happening in and around Sumer at the time has been gleaned from "Date Formulas"—royal records in which each year of the king's reign was noted by the major event that year. In the case of Shulgi much more is known, for he left behind other short and long inscriptions, including poetry and love songs.

These records indicate that soon after he had ascended the throne, Shulgi—perhaps hoping to avert his father's fate on a battlefield—reversed his father's militant policies. He launched an expedition to the outlying provinces, including the "rebel lands," but his "weapons" were offers of trade, peace, and his daughters in marriage. Deeming himself a successor to Gilgamesh, his route embraced the two destinations of that famed hero: the Sinai peninsula (where the spaceport was) in the south and the Landing Place in the north. Observing the sanctity of the Fourth Region, Shulgi skirted the peninsula and paid homage to the gods at its boundary, at a place described as "Great fortified place of the gods." Moving northward west of the Dead Sea, he paused to worship at the "Place of Bright Oracles"—the place we know as Jerusalem—and built there an altar to "the god who judges" (usually an epithet of Utu/Shamash). At the "Snow-covered Place" in the north, he built an altar and offered sacrifices. Having thus "touched base" with the reachable space-related sites, he followed the "Fertile Crescent"—the arching trade and migration east–west route dictated by geography and water sources—then continued southward in the Tigris-Euphrates plain, back to southern Sumer.

When Shulgi returned to Ur, he had every reason to think that he had brought to gods and people alike "Peace in our time" (to use a modern analogy). He was granted by the gods the title "High Priest of Anu, Priest of Nannar." He was befriended by Utu/Shamash, and was given the personal attention of Inanna/Ishtar (boasting in his love songs that she granted him her vulva in her temple).

But while Shulgi turned from affairs of state to personal pleasures, the unrest in the "rebel lands" was continuing. Unprepared for military action, Shulgi asked his Elamite ally for troops, offering its king as a reward one of his daughters in marriage and the Sumerian city Larsa as dowry. A major military expedition, employing those Elamite troops, was launched against the "sinning cities" in the west; the troops reached the Fortified Place of the gods at the Fourth Region's boundary. Shulgi in his inscriptions boasted of victory, but in fact, soon thereafter, he started to build a fortified wall to protect Sumer against foreign incursions from the west and from the northwest.

The Date Formulas called it the Great West Wall, and scholars believe that it ran from the Euphrates to the Tigris rivers north of where Baghdad is situated nowadays, blocking to invaders the way down the fertile plain between the two rivers. It was a defensive measure that preceded the Great Wall of China, which was built for similar reasons, by almost two thousand years!

In 2048 B.C.E. the gods, led by Enlil, had enough of Shulgi's state failures and personal *dolce vita*. Determining that "the divine regulations he did not carry out," they decreed for him "the death of a sinner." We don't know what kind of death it was, but it is a historic fact that in that year he was replaced on the throne of Ur by his son Amar-Sin, of whom we know from the inscriptions that he launched one military expedition after another—to quell a revolt in the north, to fight an alliance of five kings in the west.

As in so much else, what was happening had root causes going back, sometimes way back, to earlier times and events. The "rebel lands," though in Asia and thus domains in the Enlilite Lands of Noah's son Shem, were inhabited by varied "Canaanites"—offspring of the biblical Canaan who, though descended of Ham (and thus belonging to Africa), occupied a stretch of Shem's lands (*Genesis,* Chapter 10). That the "Lands of the West" along the Mediterranean coast were somehow disputed territory was also indicated by ancient Egyptian texts regarding the bitter contest between Horus and Seth that ended in aerial battles between them over the Sinai and the same contested lands.

It is noteworthy that in their military expeditions to subdue and punish the "rebel lands" in the west, both Ur-Nammu and Shulgi reached the Sinai peninsula, but turned away from that Fourth Region without entering it. The prize there was a place called TIL.MUN—the "Place of the Missiles"—the site of the post-Diluvial spaceport of the Anunnaki. When the Pyramid Wars ended, the sacred Fourth Region was entrusted to the neutral hands of Nin-mah (who was then renamed NIN.HAR.SAG—"Lady of the Mountain Peaks"), but actual command of the spaceport was put in the hands of Utu/Shamash (here shown in his winged dress uniform, **Fig. 29**, commanding the spaceport's "Eaglemen," **Fig. 30)**.

FIGURE 29

FIGURE 30

That, however, appeared to change as the struggle for supremacy intensified. Inexplicably, various Sumerian texts and "God Lists" started to associate Tilmun with Marduk's son, the god Ensag/Nabu. Enki was apparently involved in that, for a text dealing with the affair between Enki and Ninharsag states that the two of them decided to allocate the place to Marduk's son: "Let Ensag be the lord of Tilmun," they said.

The ancient sources indicate that from the safety of the sacred region Nabu ventured to the lands and cities along the Mediterranean coast, even to some Mediterranean islands, spreading everywhere the message of Marduk's coming supremacy. He was, thus, the enigmatic "Son-Man" of the Egyptian and the Akkadian prophecies—the Divine Son who was also a Son-Man, the son of a god and of an Earthling woman.

The Enlilites, understandably, could not accept such a situation. And so it was that when Amar-Sin ascended the throne of Ur after Shulgi, the targets and strategy of the Ur III military expeditions were changed in order to reassert Enlilite control over Tilmun, to sever the sacred region from the "rebel lands," then pry loose those lands from the influence of Nabu and Marduk by force of arms.

Starting in 2047 B.C.E., the sacred Fourth Region became a target and a pawn in the Enlilite struggle with Marduk and Nabu; and as both biblical and Mesopotamian texts reveal, the conflict erupted to *the greatest international "world war" of antiquity.* **Involving the Hebrew Abraham, that "War of the Kings" placed him in center stage of international events.**

In 2048 B.C.E. the destiny of the founder of monotheism, Abraham, and the fate of the Anunnaki god Marduk converged at a place called Harran.

Harran—"The Caravanry"—was an important trading center from time immemorial in Hatti (the land of the Hittites). It was located at the crossroads of major international trade and military land routes. Situated at the headwaters of the Euphrates River, it was also a hub for river transportation all the way downstream to Ur itself. Surrounded by fertile meadows watered by the river's tributaries, the Balikh and Khabur rivers, it was a center of sheepherding. The famed "Merchants of Ur" came there for Harran's wool, and brought in exchange to distribute from there Ur's famed woolen garments. Commerce in metals, skins, leather, woods, earthenware products, and spices followed. (The Prophet Ezekiel, who was exiled from Jerusalem to the Khabur area in Babylonian times, mentioned Harran's "merchants in choice fabrics, embroidered cloaks of blue, and many-colored carpets".)

Harran (the town, by that very name, still exists in Turkey, near the border with Syria, and was visited by me in 1997) was also known in ancient times as "Ur away from Ur"; at its center stood a great temple to Nannar/Sin. In 2095 B.C.E., the year in which Shulgi took over the throne in Ur, a priest named Terah was sent from Ur to Harran to serve at that temple. He took along his family; it included his son Abram. We know about Terah, his family, and their move from Ur to Harran from the Bible:

Now these are the generationas of Terah:
Terah begot Abram, Nahor and Haran,
and Haran begot Lot.
And Haran died before his father Terah
in his land of birth, in Ur in Chaldea.
And Abram and Nahor took wives—
the wife of Abram was named Sarai
and that of Nahor's wife Milkhah . . .
And Terah took with him his son Abram
and Lot, the son of his son Haran,
and his daughter-in-law Sarai,
and went forth with them from Ur in Chaldea
by the way to Canaan;
and they reached Harran and resided there.

GENESIS 11: 27–31

It is with these verses that the Hebrew Bible begins the pivotal tale of Abraham—called at the beginning by his Sumerian name **Abram**. His father, we are told earlier, stemmed from a patriarchal line that went all the way back to Shem, the oldest son of Noah (the hero of the Deluge); all those Patriarchs enjoyed long lives—Shem to the age of 600, his son Arpakhshad to 438; and subsequent male offspring to 433, 460, 239, and 230 years. Nahor, the father of Terah, lived to age 148; and Terah himself—who fathered Abram when he was seventy years old—lived to age 205. Chapter 11 of *Genesis* explains that Arpakhshad and his descendants lived in the lands later known as Sumer and Elam and their surroundings. *So Abraham, as Abram, was a true Sumerian.*

This genealogical information alone indicates that Abraham was of a special ancestry. His Sumerian name, AB.RAM, meant "Father's Beloved," an apropriate name for a son finally born to a seventy-year-old father. The father's name, Terah, stemmed from the Sumerian epithet-name TIRHU; it designated an Oracle Priest—a priest who observed celestial signs or received oracular

74

messages from a god, and explained or conveyed them to the king. The name of Abram's wife, SARAI (later *Sarah* in Hebrew), meant "Princess"; the name of Nahor's wife, *Milkhah*, meant "Queenlike"; both suggest a royal genealogy. Since it was later revealed that Abraham's wife was his half-sister—"the daughter of my father but not of my mother," he explained—it follows that Sarai/Sarah's mother was of royal descent. The family thus belonged to Sumer's highest echelons, combining royal and priestly ancestries.

Another significant clue to identifying the family's history is the repeated reference by Abraham to himself, when he met rulers in Canaan and Egypt, as being an *Ibri*—a "Hebrew." The word stems from the root *ABoR*—to come across, to cross—so it has been assumed by biblical scholars that by that he meant that he had come across from the other side of the Euphrates River, i.e., from Mesopotamia. But I believe that the term was more specific. The name used for Sumer's "Vatican City," *Nippur*, is the Akkadian rendering of the original Sumerian name NI.IBRU, "Splendid Place of Crossing." Abram, and his descendants who are called in the Bible Hebrews, belonged to a family that identified themselves as "*Ibru*"—Nippurians. That would suggest that Terah was first a priest in Nippur, then moved to Ur, and finally to Harran, taking his family along.

By synchronizing biblical, Sumerian, and Egyptian chronologies (as detailed in *The Wars of Gods and Men*), we have arrived at the year 2123 B.C.E. as the date of Abraham's birth. The gods' decision to make Nannar/Sin's cult center Ur the capital of Sumer and to enthrone Ur-Nammu took place in 2113 B.C.E. Soon thereafter, the priesthoods of Nippur and Ur were combined for the first time; it is very likely that it was then that the Nippurian priest Tirhu moved with his family, including the ten-year-old boy Abram, to serve in Nannar's temple in Ur.

In 2095 B.C.E., when Abraham was twenty-eight and already

married, Terah was transferred to Harran, taking the family with him. It could not have been just a coincidence that it was the very same year in which Shulgi succeeded Ur-Nammu. **The emerging scenario is that the movements of this family were somehow linked to the geopolitical events of that era**. Indeed, when Abraham himself was chosen to carry out divine orders to leave Harran and rush to Cannan, *the great god Marduk took the crucial step of moving to Harran*. **It was in 2048 B.C.E. that the two moves occurred: Marduk coming to sojourn in Harran, Abraham leaving Harran for faraway Cannan.**

We know from *Genesis* that Abram was seventy-five years old, and it was thus 2048 B.C.E., that he was told by God, "Get thee out of thy country and out of thy birthplace and from thy father's house"—leave behind Sumer, Nippur, and Harran—and go "unto the land which I will show thee." As to Marduk, a long text known as the *Marduk Prophecy* that he addressed to the

FIGURE 31

76

people of Harran (clay tablet, **Fig. 31**) provides the clue confirming the fact and the time of his move to Harran: 2048 B.C.E. **There is no way the two moves could have been unrelated.**

But 2048 B.C.E. was also the very year in which the Enlilite gods decided to get rid of Shulgi, ordering for him the "death of a sinner"—a move that signaled the end of "let's try peaceful means" and a return to aggressive conflict; *and there is no way that this, too, was just a coincidence.* No, the three moves—Marduk to Harran, Abram leaving Harran for Canaan, and the removal of the decadent Shulgi—had to be interconnected: **three simultaneous and interrelated moves in the Divine Chessgame.**

They were, as we shall see, steps in the countdown to Doomsday.

The ensuing twenty-four years—from 2048 to 2024 B.C.E.— were a time of religious fervor and ferment, of international diplomacy and intrigue, of military alliances and clashing armies, of a struggle for strategic superiority. The spaceport in the Sinai peninsula, and the other space-related sites, were constantly at the core of events.

Amazingly, various written records from antiquity have survived, providing us not just with an outline of events but with great details about the battles, the strategies, the discussions, the arguments, the participants and their moves, and the crucial decisions that resulted in the most profound upheaval on Earth since the Deluge.

Augmented by the Date Formulas and varied other references, the principal sources for reconstructing those dramatic events are the relevant chapters in *Genesis*; Marduk's autobiography, known as *The Marduk Prophecy*; a group of tablets in the "Spartoli Collection" in the British Museum known as *The Khedorla'omer Texts*; and a long historical/autobiographical text dictated by the god Nergal to a trusted scribe, a text known as the *Erra Epos*. As in a movie—usually a crime thriller—in which the various eye-witnesses and principals describe the same event not exactly the same way, but from which the real story emerges, so are we able to reach the same result in this case.

Marduk's main chess move, in 2048 B.C.E., was to establish his command post in Harran. By that he took away from Nannar/Sin this vital northern crossroads and severed Sumer from the northern lands of the Hittites. Besides the military significance, the move deprived Sumer of its economically vital commercial ties. The move also enabled Nabu "to marshal his cities, toward the Great Sea to set his course." Place names in these texts suggest that the principal cities west of the Euphrates River were coming under full or partial control of the father–son team, including the all-important Landing Place.

It was into the most populated part of the Lands of the West—Canaan—that Abram/Abraham was commanded to go. He left Harran, taking his wife and nephew Lot with him. He was traveling swiftly southward, stopping only to pay homage to his God at selected sacred sites. His destination was the Negev, the dry region bordering the Sinai Peninsula.

He did not stay there long. As soon as Shulgi's successor, Amar-Sin, was enthroned in Ur in 2047 B.C.E., Abram was instructed to go to Egypt. He was at once taken to meet the reigning Pharaoh, and was provided with "sheep and oxen and asses, and male attendants and female servants, and she-asses and camels." The Bible is mum regarding the reason for this royal treatment, except to hint that the Pharaoh, being told that Sarai was Abram's sister, assumed that she was being offered to him in marriage—a step that suggsts that a treaty was discussed. That such high level international negotiations were taking place between Abram and the Egyptian king seems plausible when one realizes that the year when Abram returned to the Negev after a seven-year stay in Egypt—2040 B.C.E.—was the very same year in which the Theban princes of Upper Egypt defeated the previous Lower Egypt dynasty, launching Egypt's unified Middle Kingdom. ***Another geopolitical coincidence!***

Abram, now reinforced with manpower and camels, returned to the Negev in the nick of time, his mission now clear: to defend the Fourth Region with its spaceport. As the biblical narrative reveals, he now had with him an elite force of *Ne'arim*—a term ususally translated "Young Men"—but Mesopotamian texts used the parallel term LU.NAR ("NAR-men") to denote armed cavalrymen. It is my suggesation that Abraham, having learnt in Harran tactics from the militarily excelling Hittites, obtained in Egypt a striking force of swift camel-riding cavalrymen. His base in Canaan was again the Negev, the area bordering the Sinai Peninsula.

He did so in the nick of time, for a mighty army—legions of an alliance of Enlilite kings—was on its way not only to crush and punish the "sinning cities" that switched allegiance to "other gods," but to also capture the spaceport.

The Sumerian texts dealing with the reign of Amar-Sin, Shulgi's son and successor, inform us that in 2041 B.C.E. he launched his greatest (and last) military expedition against the Lands of the West that fell under the Marduk-Nabu spell. It entailed an invasion of unparalleled scope by an international alliance, in which not only cities of men but also strongholds of gods and their offspring were attacked.

It was, indeed, such a major and unparalleled occurrence that the Bible devoted a whole long chapter to it—*Genesis,* Chapter 14. Biblical scholars call it "The War of the Kings," for it climaxed in a great battle between an army of four "Kings of the East" and the combined forces of five "Kings of the West," and culminated in a remarkable military feat by Abraham's swift cavalrymen.

The Bible begins its report of that great international war by listing the kings and kingdoms of the East who "came and made war" in the West:

> *And it came to pass*
> *in the days of Amraphel king of Shine'ar,*
> *Ariokh king of Ellasar,*
> *Khedorla'omer king of Elam,*
> *and Tidhal the king of Goyim.*

The group of tablets named the *Khedorla'omer Texts* was first brought to scholarly attention by the Assyriologist Theophilus Pinches in a lecture at the Victoria Institute, London, in 1897. They clearly describe the same events that are the great international war of Chapter 14 of *Genesis,* though in much greater detail; it is quite possible, indeed, that those tablets served as the source for the biblical writers. Those tablets identify "Khedorla'omer king of Elam" as the Elamite king Kudur-Laghamar, who is known from historical records. "Ariokh" has been identified as ERI.AKU ("Servant of the Moon god"), who reigned in the city of Larsa (biblical "Ellasar"); and Tidhal was identified as Tud-Ghula, a vassal of the king of Elam.

There has been over the years a debate regarding the identity of "Amraphel king of Shine'ar"; suggestions ranged all the way to Hammurabi, a Babylonian king centuries later. Shine'ar was the constant biblical name for Sumer, not Babylon, so who, in the time of Abraham, was its king? I have convincingly suggested in *The Wars of Gods and Men* that the Hebrew should be read not Amra-Phel but *Amar-Phel*, from the Sumerian AMAR.PAL—a variant of AMAR.SIN—whose Date Formulas attest that he did indeed, in 2041 B.C.E., launch the War of the Kings.

That fully identified coalition, according to the Bible, was led by the Elamites—a detail corroborated by the Mesopotamian data that highlights the reemerging leading role of Ninurta in the struggle. The Bible also dates this Khedorla'omer Invasion by observing that it took place fourteen years after the previous Elamite incursion into Canaan—another detail conforming to the data from Shulgi's time.

The invasion route this time was, however, different: short-cutting the distance from Mesopotamia by a risky passage through a stretch of desert, the invaders avoided the densely populated Mediterranean coastland by marching on the eastern side of the Jordan River. The Bible lists the places where those battles took place and who the Enlilite forces battled there; the information indicates that an attempt was made to settle accounts with old adversaries—descendants of the intermarrying Igigi, even of the Usurper Zu—who evidently supported the uprisings against the Enlilites. But sight was not lost of the prime target: *the spaceport*. The invading forces followed what has been known since biblical times as the Way of the King, running north–south on the eastern side of the Jordan. But when they turned westward toward the gateway to the Sinai Peninsula, they met a blocking force: Abraham and his cavalrymen **(Fig. 32)**.

Referring to the Peninsula's gateway city Dur-Mah-Ilani ("The gods' great fortified place")—the Bible called it Kadesh-Barnea—the *Khedorla'omer Texts* clearly stated that the way was blocked there:

> *The son of the priest,*
> *whom the gods in their true counsel had anointed,*
> *the despoiling had prevented.*

"The son of the priest," *anointed by the gods*, **I suggest, was Abram the son of the priest Terah.**

A Date Formula tablet belonging to Amar-Sin, inscribed on both sides **(Fig. 33)**, boasts of destroying NE IB.RU.UM—"The Shepherding place of *Ibru'um*." In fact, at the gateway to the spaceport there was no battle; the mere presence of Abram's cavalry striking force persuaded the invaders to turn away—to richer

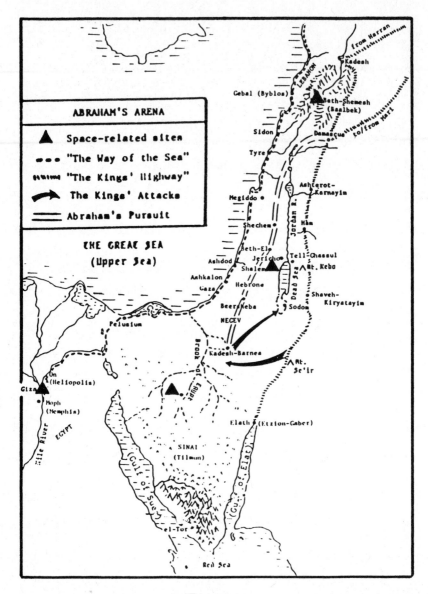

FIGURE 32

and more lucrative targets. But *if the reference is indeed to Abram, by name, it offers once more an extraordinary extra-biblical corroboration of the Patriarchal record, no matter who claimed victory.*

Prevented from entering the Sinai Peninsula, the Army of the East turned northward. The Dead Sea was then shorter; its current

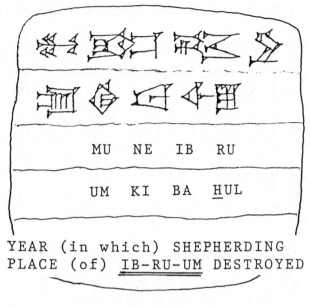

MU NE IB RU

UM KI BA HUL

YEAR (in which) SHEPHERDING
PLACE (of) IB-RU-UM DESTROYED

FIGURE 33

southern appendix was not yet submerged, and it was then a fertile plain rich with farmland, orchards, and trading centers. The settlements there included five cities, among them the infamous Sodom and Gomorrah. Turning northward, the invaders now faced the combined forces of what the Bible called "five sinning cities." It was there, the Bible reports, that the four kings fought and defeated the five kings. Looting the cities and taking captives with them, the invaders marched back, this time on the western side of the Jordan.

The biblical focus on those battles might have ended with that turning back were it not for the fact that Abram's nephew Lot, who resided in Sodom, was among the captives. When a refugee from Sodom told Abram what had happened, "he armed his trained men, three hundred and eighteen of them, and gave chase." His cavalry caught up with the invaders all the way north, near Damascus (see Fig. 32), where Lot was freed and the booty recovered. The Bible records the feat as the "smiting of Khedorla'omer and the kings who were with him" by Abram.

The historical records suggest that as audacious and far-flung that War of the Kings had been, it failed to suppress the Marduk-Nabu surge. Amar-Sin, we know, died in 2039 B.C.E.—felled not by an enemy lance, but by a scorpion's bite. He was replaced in 2038 B.C.E. by his brother Shu-Sin. The data for his nine years' reign record two military forays northward but none westward; they speak mostly of his defensive measures. He relied mainly on building new sections of the Wall of the West against attacking Amorites. The defenses, however, were moved each time ever closer to Sumer's heartland, and the territory controlled from Ur kept shrinking.

By the time the next (and last) of the Ur III dynasty, Ibbi-Sin, ascended the throne, invaders from the west had broken through the defensive Wall and were clashing with Ur's "Foreign Legion," Elamite troops, in Sumerian territory. Directing and prompting the Westerners on toward the cherished target was Nabu. His divine father, Marduk himself, was waiting in Harran for the recapture of Babylon.

The great gods, called to an emergency council, then approved extraordinary steps that changed the future forever.

Chapter VI

GONE WITH THE WIND

The unleashing of "weapons of mass destruction" in the Middle East underlies the fear of Armageddon prophecies coming true. The sad fact is that a mounting conflict—among gods, not men—did lead to the use of nuclear weapons, right there, four thousand years ago. And if there ever was a most regrettable act with the most unexpected consequences, that was it.

That nuclear weapons had been used on Earth for the first time not in 1945 A.D. but in 2024 B.C.E. is fact, not fiction. The fateful event is described in a variety of ancient texts from which the What and How, the Why and Who can be construed, reconstructed and put in context. Those ancient sources include the Hebrew Bible, for the first Hebrew Patriarch, Abraham, was an eyewitness to the awesome calamity.

The failure of the War of the Kings to subdue the "rebel lands" of course discouraged the Enlilites and encourged the Mardukites, but the events did more than that. On Enlil's instructions, Ninurta got busy setting up an alternative space facility on the other side of the world—all the way in what is now Peru in South America. The texts indicate that Enlil himself was away from Sumer for long stretches of time. These gods' moves caused the last two kings of Sumer, Shu-Sin and Ibbi-Sin, to waver in their allegiances and to start paying homage to Enki in his Sumerian foothold, Eridu. The divine absences also loosened controls over the Elamite "Foreign Legion," and the records speak of "sacrileges" by the Elamite troops. Gods and men were increasingly disgusted with it all.

Especially enraged was Marduk, who received word of looting, destructions, and desecrations in his cherished Babylon. It will be recalled that the last time he was there he was persuaded by his half-brother Nergal to leave peacefully until the Celestial Time would reach the Age of the Ram. He did so having received Nergal's solemn word that nothing would be disturbed or desecrated in Babylon, but the opposite happened. Marduk was angered by the reported desecration of his temple there by the "unworthy" Elamites: "To herds of dogs Babylon's temple they made a den; flying ravens, loudly shrieking, their dung dropped there."

From Harran he cried out to the great gods: "Until When?" Has not the Time arrived yet, he asked in his prophetic autobiography:

> O great gods, learn my secrets
> as I girdle my belt, my memories remember.
> I am the divine Marduk, a great god.
> I was cast off for my sins,
> to the mountains I have gone.
> In many lands I have been a wanderer.
> From where the sun rises to where it sets I went.
> To the highland of Hatti I came.
> In Hattiland I asked for an oracle;
> in it I asked: "Until when?"

"Twenty-four years in Harran's midst I nested," Marduk went on; "my days are completed!" The time has come, he said, to set his course to his city (Babylon), "my temple to rebuild, my ever-lasting abode to establish." Waxing visionary, he spoke of seeing his temple E.SAG.ILA ("Temple whose head is lofty") rising as a mountain upon a platform in Babylon, calling it "The house of my covenant." He foresaw Babylon as forever established, a king of his choice installed there, a city filled with joy, a city blessed by Anu. The messianic times, Marduk prophesied, will "chase away evil and bad luck, bring motherly love to Mankind."

The year in which a sojourn of twenty-four years in Harran was completed was 2024 B.C.E.; it marked seventy-two years since Marduk had agreed to depart from Babylon and await the oracular celestial time.

Marduk's "until when?" appeal to the Great Gods was not an idle one, for the leadership of the Anunnaki was constantly con-sulting, informally and in formal councils. Alarmed by the wors-ening situation, Enlil hurriedly returned to Sumer, and was shocked to learn that things had gone wrong even in Nippur it-self. Ninurta was summoned to explain the Elamites' miscon-duct, but Ninurta put all the blame on Marduk and Nabu. Nabu was summoned, and "Before the gods the son of his father came." His main accuser was Utu/Shamash, who, describing the dire situation, said, "all this Nabu has caused to happen." Speaking for his father, Nabu blamed Ninurta, and revived the old accusations against Nergal in regard to the disappearance of the pre-Diluvial monitoring instruments and the failure to prevent sacrileges in Babylon; he got into a shouting match with Nergal, and "show-ing disrespect . . . to Enlil evil he spoke: 'There is no justice, de-struction was conceived, Enlil against Babylon caused evil to be planned.'" It was an unheard-of accusation against the Lord of the Command.

Enki spoke up, but it was in defense of his son, not of Enlil. What are Marduk and Nabu actually accused of? he asked. His ire was directed especially at his son Nergal: "Why do you con-tinue the opposition?" he asked him. The two argued so much that in the end Enki shouted to Nergal to get out of his presence. The gods' councils broke up in disarray.

But all these debates, accusations, and counteraccusations were taking place against the increasingly realized fact—what Marduk referred to as the Celestial Oracle: with the passage of time—with the crucial shift of the precessional clock by one degree—the Age of the Bull, the zodiacal age of Enlil, was coming to an end, and the Age of the Ram, Marduk's Age, was looming in the heavens. Ninurta could see it coming at his Eninnu temple in Lagash (which Gudea built); Ningishzidda/Thoth could confirm it from all the stone circles that he had erected elsewhere on Earth; and the people knew it, too.

It was then that Nergal—vilified by Marduk and Nabu, ordered out by his father Enki—"consulting with himself," concocted the idea of resort to the "Awesome Weapons." He did not know where they were hidden, but knew that they existed on Earth, locked away in a secret underground place (according to a text catalogued as CT-xvi, lines 44–46, somewhere in Africa, in the domain of his brother Gibil):

> *Those seven, in the mountains they abide;*
> *In a cavity inside the earth they dwell.*

Based on our current level of technology, they can be described as seven nuclear devices: "Clad with terror, with a brilliance they rush forth." They were brought to Earth unintentionally from Nibiru and were hidden away in a secret safe place a long time ago; Enki knew where, but so did Enlil.

A War Council of the gods, overruling Enki, voted to follow Nergal's suggestion to give Marduk a punishing blow. There was constant communication with Anu: "Anu to Earth the words was speaking, Earth to Anu the words pronounced." He made it clear that his approval for the unprecedented step was limited to depriving Marduk of the Sinai spaceport, but that neither gods nor people should be harmed: "Anu, lord of the gods, on the Earth had pity," the ancient records state. Choosing Nergal and Ninurta to carry out the mission, the gods made absolutely clear to them its limited and conditional scope.

But that is not what happened: The "Law of Unintended Consequences" proved itself true on a catastrophic scale.

In the aftermath of the calamity that resulted in the death of countless people and the desolation of Sumer, Nergal dictated to a trusted scribe his own version of the events, trying to exonerate himself. The long text is known as the *Erra Epos,* for it refers to Nergal by the epithet *Erra* ("The Annihilator") and to Ninurta as *Ishum* ("The Scorcher"). We can put together the true story by adding to this text information from several other Sumerian, Akkadian, and biblical sources.

Thus we find that no sooner was the decision reached than Nergal rushed to Gibil's African domain to find and retrieve the weapons, not waiting for Ninurta. To his dismay Ninurta learnt that Nergal was disregarding the objective's limits, and was going to use the weapons indiscriminately to settle personal accounts: "I shall annihilate the son, and let the father bury him; then I shall kill the father, and let no one bury him," Nergal has boasted.

While the two argued, word reached them that Nabu was not sitting still: "From his temple to marshall all his cities he set his step, toward the Great Sea he set his course; the Great Sea he entered, sat upon a throne that was not his." Nabu was not only converting the western cities, he was taking over the Mediterranean islands, and setting himself up as their ruler! Nergal/ Erra thus argued that destroying the spaceport was not enough: Nabu, and the cities that rallied to him, also had to be punished, destroyed!

Now, with two targets, the Nergal-Ninurta team saw another

problem: Would the "upheavaling" of the spaceport not sound the alarm for Nabu and his sinning followers to escape? Reviewing their targets, they found the solution in splitting up: Ninurta would attack the spaceport; Nergal would attack the nearby "sinning cities." But as all this was agreed upon, Ninurta had second thoughts; he insisted that not only the Anunnaki who manned the space facilities should be forewarned, but that even certain people should be forewarned: "Valiant Erra," he told Nergal, *"will you the righteous destroy with the unrighteous? Will you destroy those who against you have not sinned with those who against you have sinned?"*

Nergal/Erra, the ancient text states, was persuaded: "The words of Ishum appealed to Erra as fine oil." And so, one morning, the two, sharing the seven nuclear explosives between them, set out on their ultimate Mission:

> *Then did the hero Erra go ahead,*
> *remembering the words of Ishum.*
> *Ishum too went forth*
> *in accordance with the words given,*
> *a squeezing in his heart.*

The available texts even tell us who went to what target: "Ishum to the Mount Most Supreme set his course" (we know that the spaceport was beside this mount from the Epic of Gilgamesh). "Ishum raised his hand: the Mount was smashed . . . That which was raised toward Anu to launch was caused to wither, its face was made to fade away, its place was made desolate." In one nuclear blow, the spaceport and its facilities were obliterated by the hand of Ninurta.

The ancient text then describes what Nergal did: "Emulating Ishum, Erra the Way of the King followed, the cities he finished off, to desolation he overturned them"; his targets were the "sinning cities" whose kings had formed the alliance against the Kings of the East, the plain in the south of the Dead Sea.

And so it was that in the year 2024 B.C.E. nuclear weapons were unleashed in the Sinai Peninsula and in the nearby Plain of the Dead Sea; and the spaceport and the Five Cities were no more.

Amazingly, yet no wonder if Abraham and his mission in Canaan is understood the way we explain it, it is in this apocalyptic event that the biblical record and the Mesopotamian texts converge.

We know from the Mesopotamian texts relating the events that, as required, the Anunnaki guarding the spaceport were forewarned: "The two [Nergal and Ninurta], incited to commit the evil, made its guardians stand aside; the gods of that place abandoned it—its protectors went up to the heights of heaven." But while the Mesopotamian texts reiterate that "the two made the gods flee, made them flee the scorching," they are ambiguous regarding whether that advance notice was also extended to the people in the doomed cities. It is here that the Bible provides missing details: we read in *Genesis* that both Abraham and his nephew Lot were indeed forewarned—but not the other residents of the "sinning cities."

The biblical report, apart from throwing light on the "upheavaling" aspects of the events, contains details that shed an amazing light on the gods in general and on their relationship with Abraham in particular. The story begins in Chapter 18 of *Genesis* when Abraham, now ninety nine years old, sitting at the entrance to his tent on a hot midday, "lifted his eyes" and all of a sudden saw "three men standing above him." Though they are described as *Anashim,* "men," there was something different or unusual about them, for he rushed out of his tent and bowed to the ground, and—referring to himself as their servant—washed their feet and offered them food. As it turned out, the three were divine beings.

As they leave, their leader—now identified as the Lord God—decides to reveal to Abraham the trio's mission: to determine whether Sodom and Gomorrah are indeed sinning cities whose upheavaling is justified. While two of the three continue toward Sodom, Abraham approaches and **reproaches** (!) God with words that are identical to those in the Mesopotamian text: *Wilt thou destroy the righteous with the unrighteous?* (*Genesis* 18: 23).

What followed was an incredible bargaining session between Man and God. "Perchance there are fifty righteous within the city—Wilt thou destroy, and not spare the city on account of the fifty righteous within it?" Abraham asked God. When told that, well, the city would be spared if fifty righteous men reside there, Abraham said, what about just forty? What about only thirty? And so it went, down to ten . . . "And Yahweh went away as soon as he had finished speaking, and Abraham returned to his place."

The other two divine beings—the tale's continuation in Chapter 19 calls them *Mal'achim*, literally "emissaries" but commonly translated "Angels"—arrived in Sodom in the evening. The happenings there confirmed its people's wickedness, and at daybreak the two urged Abraham's nephew Lot to escape with his family, for "Yahweh is about to destroy the city." The lingering family asked for more time, and one of the "angels" agreed to have the upheaval delayed long enough for Lot and his family to reach the safer mountain.

"And Abraham got up early in the morning . . . and he looked toward Sodom and Gomorrah and toward all the land of the Plain, and beheld, and lo—vapor went up from the earth as the smoke of a furnace."

Abraham was then ninety-nine years old; having been born in 2123 B.C.E., the time had to be 2024 B.C.E.

The convergence of the Mesopotamian texts with the biblical narrative of *Genesis* concerning the upheaval of Sodom and Gomorrah is at once one of the most significant confirmations of the Bible's veracity in general and of Abraham's status and role in particular—and yet one of the most shunned by theologians and other scholars, because of its report of the events of the preceding day, the day three Divine Beings ("Angels" who looked like men) had paid Abraham a visit—it smacks too much of an "Ancient Astronauts" tale. Those who question the Bible or treat the Mesopotamian texts as just myths have sought to explain the destruction of Sodom and Gomorrah as some natural calamity, yet the biblical version confirms twice that the "upheaval" by "fire and sulfur" was not a natural calamity but *a premeditated, postponable and even cancellable* event: once when Abraham bargained with The Lord to spare the cities so as not to destroy the righteous with the unjust, and again when his nephew Lot obtained a postponement of the upheaval.

Photographs of the Sinai Peninsula from space **(Fig. 34)** still show the immense cavity and the crack in the surface where the nuclear explosion had taken place. The area itself is strewn, to

FIGURE 34

this day, with crushed, burnt, and blackened rocks **(Fig. 35)**; they contain a highly unusual ratio of isotope uranium-235, indicating in expert opinions exposure to ***sudden immense heat of nuclear origin***.

The upheaval of the cities in the plain of the Dead Sea caused

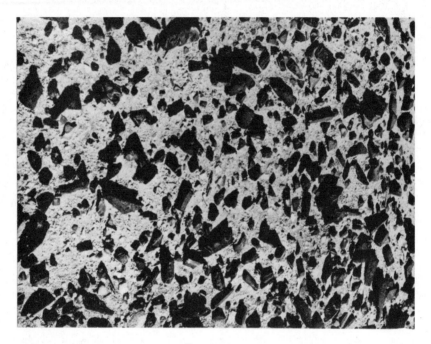

FIGURE 35

the southern shore of the sea to collapse, leading to a flooding of the once fertile area and its appearance, to this day, as an appendage separated from the sea by a barrier called *"El-Lissan"* ("The Tongue") **(Fig. 36)**. Attempts by Israeli archaeologists to explore the seabed there have revealed the existence of enigmatic underwater ruins, but the Hashemite Kingdom of Jordan, in whose half of the Dead Sea the ruins are, put a stop to further exploration. Interestingly, the relevant Mesopotamian texts confirm the topographic change and even suggest that the sea became a Dead

FIGURE 36

Sea as a result of the nuclear bombing: Erra, they tell, "Dug through the sea, its wholeness he divided; that which lives in it, even the crocodiles, he made wither."

The two, as it turned out, did more than destroy the space-port and the sinning cities: as a result of the nuclear explosions, sions,

A storm, the Evil Wind,
went around in the skies.

And the chain reaction of unintended consequences began.

The historical records show that the Sumerian civilization collapsed in the sixth year of the reign in Ur of Ibbi-Sin—in 2024 B.C.E. It was, the reader wll recall, the very year in which Abraham was ninety-nine years old . . .

Scholars assumed at first that Sumer's capital, Ur, was overrun by "barbarian invaders"; but no evidence for such a destructive invasion was found. A text titled "A Lamentation Over the Destruction of Ur" was then discovered; it puzzled the scholars, for it bewailed not the physical destruction of Ur but its "abandonment": the gods who had dwelt there abandoned it, the people who dwelt there were gone, its stables were empty; the temples, the houses, the sheepfolds remained intact—standing, but empty.

Other lamentation texts were then discovered. They lamented not just Ur, but all of Sumer. Again they spoke of "abandonment": not only did the gods of Ur, Nannar, and Ningal abandon Ur; Enlil, "the wild bull," abandoned his beloved temple in Nippur; his spouse Ninlil was also gone. Ninmah abandoned her city Kesh; Inanna, "the queen of Erech," abandoned Erech; Ninurta forsook his temple Eninnu; his spouse Bau was also gone from Lagash. One Sumerian city after another was listed as having been "abandoned," without their gods, people, or animals. The scholars were now puzzling over some "dire catastrophe," a mysterious calamity that affected the whole of Sumer. What could it be?

The answer to the puzzle was right there in those texts: *Gone with the wind.*

No, this is not a play of words on the title of a famous book/movie. That was the refrain in the Lamentation Texts: Enlil has abandoned his temple, he was "gone by the wind." Ninlil from her temple was "gone by the wind." Nannar has abandoned Ur—his sheepfolds were "gone by the wind"; and so on and on. The scholars have assumed that this repetition of the words was a literary device, a refrain that the lamenters repeated over and over again to highlight their grief. But that was no literary device—that was the literal truth: **Sumer and its cities were literally emptied as a result of a wind.**

An "*Evil Wind*," the lamentation (and then other texts) reported, came blowing and caused "a calamity, one unknown to men, to befall the land." It was an *Evil Wind* that "caused cities to be desolate, caused houses to be desolate, caused stalls to be desolate, the sheepfolds to be emptied." There was desolation, but no destruction; emptiness, but no ruins: the cities were there, the houses were there, the stalls and sheepfolds were there—but nothing alive remained; even "Sumer's rivers flow with water that is bitter, the once cultivated fields grow weeds, in the meadows the plants have withered." All life is gone. It was a calamity that had never happened before—

> On the Land Sumer a calamity fell,
> One unknown to men.
> One that had never been seen before,
> One which could not be withstood.

Carried by the Evil Wind, it was a death from which there was no escape: it was a death "which roams the street, is let loose in the road . . . The highest wall, the thickest wall, it passes like a flood; no door can shut it out, no bolt can turn it back." Those who hid behind doors were felled inside; those who ran to the rooftops died on the roofs. It was an unseen death: "It stands beside a man, yet no one can see it; when it enters a house, its appearance is unknown." It was a gruesome death: "Cough and phlegm weakened the chest, the mouth was filled with spittle, dumbness and daze have come upon them . . . an overwhelming dumbness . . . a headache." As the Evil Wind clutched its victims, "their mouths were drenched with blood." The dead and dying were everywhere.

The texts make clear that the Evil Wind, "bearing gloom from city to city," was not a natural calamity; it resulted from a deliberate decision of the great gods. It was caused by "a great storm ordered by Anu, a [decision] from the heart of Enlil." And it was the result of a single event—"spawned in a single spawning, in a lightning flash"—an event that occurred far away in the west: "From the midst of the mountains it had come, from the Plain of No-Pity it had come . . . Like a bitter venom of the gods, from the west it had come."

That the cause of the Evil Wind was the nuclear "upheaval" back in and near the Sinai peninsula was made clear when the texts asserted that the gods knew its source and cause—*a blast, an explosion*:

> *An evil blast heralded the baleful storm,*
> *An evil blast was its forerunner.*
> *Mighty offspring, valiant sons,*
> *were the heralds of the pestilence.*

The authors of the lamentation texts, the gods themselves, left us a vivid record of what had taken place. As soon as the Awesome Weapons were launched from the skies by Ninurta and Nergal, "they spread awesome rays, scorching everything like fire." The resulting storm "in a flash of lightning was created." A "dense cloud that brings doom"—a nuclear "mushroom"—then rose to the sky, followed by "rushing wind gusts . . . a tempest that scorches the heavens." It was a day not to be forgotten:

> *On that day,*
> *When heaven was crushed*

and the Earth was smitten,
its face obliterated by the maelstrom—
When the skies were darkened
and covered as with a shadow—
On that day the Evil Wind was born.

The various texts kept attributing the venomous maelstrom to the explosion at the "place where the gods ascend and descend"—to the obliteration of the spaceport, rather than to the destruction of the "sinning cities." It was there, "in the midst of the mountains," that the nuclear mushroom cloud arose in a brilliant flash—and it was from there that the prevailing winds, coming from the Mediterranean Sea, carried the poisonous nuclear cloud eastward, toward Sumer, and there it caused not destruction but a silent annihilation, bringing death by nuclear poisoned air to all that lives.

It is evident from all the relevant texts that, with the possible exception of Enki, who had protested and warned against the use of the Awesome Weapons, none of the gods involved expected the eventual outcome. Most of them were Earthborn, and to them the tales of the nuclear wars on Nibiru were Tales of the Elders. Did Anu, who should have known better, think perhaps that the weapons, hidden so long ago, would hardly work or not work at all? Did Enlil and Ninurta (who had come from Nibiru) assume that the winds, if at all, would blow the nuclear cloud toward the desolate deserts that are now Arabia? There is no satisfactory answer; the texts only state that "the great gods paled at the storm's immensity." But it is clear that as soon as the direction of the winds and the intensity of the nuclear venom were realized, an alarm was sounded for those in the wind's path—gods and people alike—to run for their lives.

The panic, fear, and confusion that overtook Sumer and its cities as the alarm was sounded are vividly described in a series of lamentation texts, such as the *Ur Lamentation*, the *Lamentation over*

THE END OF DAYS

the Desolation of Ur and Sumer, The Nippur Lamentation, The Uruk Lamentation, and others. As far as the gods were concerned, it appears that it was by and large "each man for himself"; using their varied craft, they took off by air and by water to get out of the wind's path. As for the people, the gods did sound the alarm before they fled. As described in The Uruk Lamentation, "Rise up! Run away! Hide in the steppe!" the people were told in the middle of the night. "Seized with terror, the loyal citizens of Uruk" ran for their lives, but they were felled by the Evil Wind anyway.

The picture, though, was not identical everywhere. In Ur, the capital, Nannar/Sin was so incredulous that he refused to believe that Ur's fate has been sealed. His long and emotional appeal to his father Enlil to avert the calamity is recorded in the Ur Lamentation (which was composed by Ningal, Nannar's spouse); so is Enlil's blunt admission of inevitability:

> Ur was granted kingship—
> An eternal reign it was not granted . . .

Unwilling to accept the inevitable and too devoted to the people of Ur to abandon them, Nannar and Ningal decided to stay put. It was daytime when the Evil Wind approached Ur; "of that day I still tremble," Ningal wrote, "but of that day's foul smell we did not flee." As doomsday came, "a bitter lament was raised in Ur, but of its foulness we did not flee." The divine couple spent the night of nightmares in the "termite house," an underground chamber deep inside their ziggurat. By morning, as the venomous wind "was carried off from the city," Ningal realized that Nannar was ill. She hastily put on garments and had the god carried out and away from Ur, the city that they loved.

At least another deity was also harmed by the Evil Wind; she was Ninurta's spouse Bau, who was alone in Lagash (for her husband was busy destroying the spaceport). Loved by the people, who called her "Mother Bau," she was trained as a healing physician, and just could not force herself to leave. The lamentations record that "On that day, the storm caught up with the Lady Bau; as if she was a mortal, the storm caught up with her." It is not clear how badly she was stricken, but subsequent records from Sumer suggest that she did not survive long thereafter.

Eridu, Enki's city, lying farthest to the south, was apparently at the edge of the Evil Wind's path. We learn from *The Eridu Lament* that Ninki, Enki's spouse, flew away from the city to a safe haven in Enki's African Abzu: "Ninki, the Great Lady, flying like a bird, left her city." But Enki himself departed from the city only far enough to get out of the Evil Wind's way: "The Lord of Eridu stayed outside his city . . . for the fate of his city he wept with bitter tears." Many of Eridu's citizens followed him, camping in the fields at a safe distance as they watched—for a day and a half—the storm "put its hand on Eridu."

Amazingly, the least affected of all the land's major centers was Babylon, for it lay beyond the storm's northern edge. As the alert was sounded, Marduk contacted his father to seek advice: What are the people of Babylon to do? he asked. Those who can escape should go north, Enki told him; and in the manner of the two "Angels" who had advised Lot and his family not to look back when they fled Sodom, so did Enki instruct Marduk to tell his followers "neither to turn nor to look back." If escape was not possible, the people should seek shelter underground: "Get them into a chamber below the earth, into a darkness," was Enki's advice. Following this advice, and due to the wind's direction, Babylon and its people were unharmed.

As the Evil Wind passed and blew away (its remnants, we learn, reached the Zagros Mountains farther east), it left Sumer desolate and prostrate. "The storm desolated the cities, desolated the houses." The dead, lying where they fell, remained unburied: "The dead people, like fat placed in the sun, of themselves melted away." In the grazing lands, "cattle large and small became scarce, all living creatures came to an end." The sheepfolds "were delivered to the Wind." The cultivated fields withered; "on the banks of the Tigris and the Euphrates only sickly weeds grew, in the swamps the reeds rotted in a stench." "No one treads the highways, no one seeks out the roads."

"Oh Temple of Nannar in Ur, bitter is thy desolation!" the lamentation poems bewailed; "Oh Ningal whose land has perished, make thy heart like water!"

> *The city has become a strange city,*
> *how can one now exist?*
> *The house has become a house of tears,*
> *it makes my heart like water.*
> *Ur and its temples have been*
> *delivered to the Wind.*

After two thousand years, the great Sumerian civilization was gone with the wind.

In recent years archaeologists have been joined by geologists, climatologists, and other earth sciences experts for multidisciplinary efforts to tackle the enigma of the abrupt collapse of Sumer & Akkad at the end of the third millenium B.C.E.

A trend-setting study was one by an international group of seven scientists from different disciplines titled "Climate Change and the Collapse of the Akkadian Empire: Evidence from the Deep Sea," published in the scientific journal *Geology* in its April 2000 issue. Their research used radiological and chemical analysis of ancient dust layers from that period obtained from various Near Eastern sites, but primarily from the bottom of the Gulf of Oman; their conclusion was that an unusual climate change *in the areas adjoining the Dead Sea* gave rise to dust storms and that the dust—an unusual "atmospheric mineral dust"—was carried by the prevailing winds over southern Mesopotamia all the way beyond the

Persian Gulf **(Fig. 37)**—the very pattern of Sumer's Evil Wind! Carbon dating of the unusual "fallout dust" led to the conclusion that it was due to an *"uncommon dramatic event that occurred near 4025 years before the present."* **That, in other words, means "near 2025 B.C.E."**—*the very 2024 B.C.E. indicated by us!*

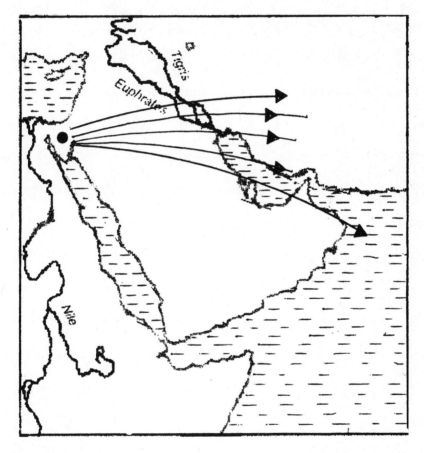

FIGURE 37

Interestingly, the scientists involved in that study observed in their report that *"the Dead Sea level fell abruptly by 100 meters at that time."* They leave the point unexplained—but obviously the breach of the Dead Sea's southern barrier and the flooding of the Plain, as described by us, explain what had happened.

The scientific journal *Science* devoted its issue of 27 April 2001 to Paleoclimate worldwide. In a section dealing with the events in Mesopotamia, it refers to evidence from Iraq, Kuwait, and Syria that the "widespread abandonment of the alluvial plain" between the Tigris and Euphrates rivers was due to dust storms "commencing 4025 years B.P." ("Before the Present"). The study leaves unexplained the cause of the abrupt "climate change," but it adopts the same date for it: 4025 years before A.D. 2001.

The fateful year, modern science confirms, was 2024 B.C.E.

Chapter VII

DESTINY HAD FIFTY NAMES

The resort to nuclear weapons at the end of the twenty-first century B.C.E. ushered—one could say, "with a bang"—the Era of Marduk. It was, in almost all respects, truly a New Age, even the way we understand the term nowadays. Its greatest paradox was that while it made Man look to the heavens, it brought the gods of the heavens down to Earth. The changes that New Age has wrought affect us to this day.

For Marduk the New Age was a wrong righted, an ambition attained, prophecies fulfilled. The price paid—the desolation of Sumer, the flight of its gods, the decimation of its people—was not his doing. If anything, those who suffered were punished for obstructing Destiny. The unforeseen nuclear storm, the Evil Wind, and its course that seemed selectively guided by an unseen hand only confirmed what the Heavens proclaimed: **the Age of Marduk, the Age of the Ram, has arrived.**

The change from the Age of the Bull to the Age of the Ram was especially celebrated and marked in Marduk's homeland, Egypt. Astronomical depictions of the heavens (such as at the Denderah temple, see Fig. 20) showed the constellation of the Ram as

1. Aries.

2. Taurus.

3. Gemini.

4. Cancer.

5. Leo

6. Virgo.

7. Libra.

8. Scorpio.

9. Sagittarius.

10. Capricorn.

11. Aquarius.

12. Pisces.

FIGURE 38

FIGURE 39

the focal point of the zodiacal cycle. Lists of zodiacal constellations began not with the Bull as in Sumer, but with the Ram **(Fig. 38)**. The most impressive manifestations were the rows of Ram-headed sphinxes that flanked the processional way to the great temples in Karnak **(Fig. 39)**, whose construction, by Pharaohs of the newly established Middle Kingdom, began right after Ra/Marduk's ascent to supremacy. They were Pharaohs who bore theophoric names honoring Amon/Amen, so that both temples and kings were dedicated to Marduk/Ra as **Amon,** *The Unseen,* for Marduk, absenting himself from Egypt, selected Babylon in Mesopotamia to be his Eternal City.

Both Marduk and Nabu survived the nuclear maelstrom unharmed. Although Nabu was personally targeted by Nergal/ Erra, he apparently hid on one of the Mediterranean islands and escaped harm. Subsequent texts indicate that he was given his own cult center in Mesopotamia called Borsippa, a new city situated near his father's Babylon, but he continued to roam and be worshipped in his favorite Lands of the West. His veneration both there and in Mesopotamia is attested to by sacred places named in his honor—such as Mount Nebo near the Jordan River (where Moses later died)—and the theophoric royal names (such as Nabo-pol-assar, Nebo-chad-nezzar, and many others) by which famous kings of Babylon were called. And his name, as we have noted, became synonymous with "prophet" and prophecy throughout the ancient Near East.

Marduk himself, it will be recalled, was asking "Until when?" from his command post in Harran when the fateful events took place. In his autobiographical text *The Marduk Prophecy* he envisioned the **coming of a Messianic Time**, when gods and men will recognize his supremacy, when peace shall replace war and abundance will banish suffering, when a king of his choice "will make Babylon the foremost" with the *Esagil* temple (as its name meant) raising its head to heaven—

> *A king in Babylon will arise;*
> *In my city Babylon, in its midst,*
> *my temple to heaven he will raise;*
> *The mountainlike Esagil he will renew,*
> *the ground plan of Heaven-Earth*
> *for the mountainlike Esagil he will draw;*
> *The Gate of Heaven will be opened.*
> *In my city Babylon a king will arise;*
> *In abundance he will reside;*
> *My hand he will grasp,*
> *He will lead me in processions . . .*
> *To my city and my temple Esagil*
> *for eternity I shall enter.*

That new Tower of Babel, however, was not intended (as the first one was) as a launch tower. His supremacy, Marduk recognized, was now stemming not only from the possession of a physical space connection but from the Signs of Heaven—from the zodiacal Celestial Time, from the position and movement of the celestial bodies, the *Kakkabu* (stars/planets) of heaven.

Accordingly, he envisioned the future Esagil as the reigning astronomical observatory, making redundant Ninurta's Eninnu and the varied stonehenges erected by Thoth. When the Esagill was eventually built, it was a ziggurat erected according to detailed and precise plans **(Fig. 40)**: its height, the spacing of its seven stages, and its orientation were such that its head pointed directly to the star *Iku*—the lead star of the constellation of the Ram—circa **1960 B.C.E.**

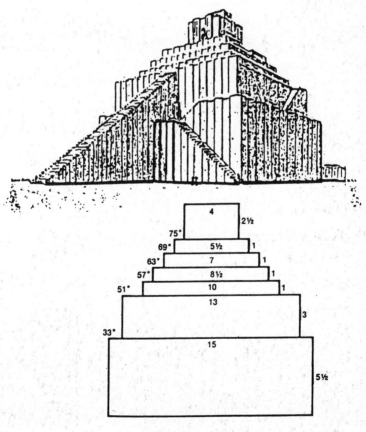

FIGURE 40

The nuclear apocalypse and its unintended consequences brought to an abrupt end the debate regarding whose zodiacal age it was; Celestial Time was now Marduk's Time. But the gods' planet, Nibiru, was still orbiting and clocking Divine Time—and Marduk's attention shifted to that. As his Prophecy text made clear, he now envisioned astronomer-priests scanning the skies from the ziggurat's stages for **"The rightful planet of the Esagil"**:

> *Omen-knowers, put to service.*
> *shall then ascend its midst.*
> *Left and right, on opposite sides,*
> *they shall separately stand.*
> *The king will then approach;*
> *The rightful* Kakkabu *of the Esagil*
> *over the land [he will observe].*

A Star-Religion was born. The god—Marduk—became a star; a star (we call it planet)—Nibiru—became "Marduk." Religion became Astronomy, Astronomy became Astrology.

In conformity with the new Star Religion, the Epic of Creation, *Enuma Elish,* was revised in its Babylonian version so as to grant Marduk a celestial dimension: he did not just come from Nibiru—he *was* Nibiru. Written in "Babylonian," a dialect of Akkadian (the Semitic mother language), it equated Marduk with Nibiru, the home planet of the Anunnaki, and gave the name "Marduk" to the Great Star/Planet that had come from deep space to avenge both the celestial Ea and the one on Earth **(Fig. 41)**. It thus made "Marduk" the "Lord" in Heaven as on Earth. His Destiny—in the heavens, his orbit—was the greatest of all the celestial gods (the other planets) (see Fig. 1); paralleling that, he was destined to be the greatest of the Anunnaki gods on Earth.

The revised Epic of Creation was read publicly on the fourth

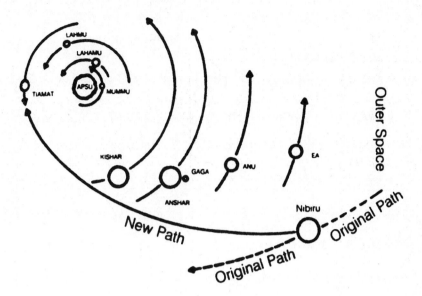

FIGURE 41

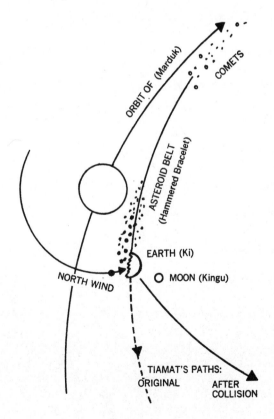

FIGURE 42

night of the New Year festival. It credited Marduk with the defeat of the "monster" Tiamat in the Celestial Battle, the creation of the Earth **(Fig. 42)**, and the reshaping of the Solar system **(Fig. 43)**—all the feats that in the original Sumerian version were attributed to the planet Nibiru as part of a sophisticated scientific cosmogony. The new version then credited Marduk even

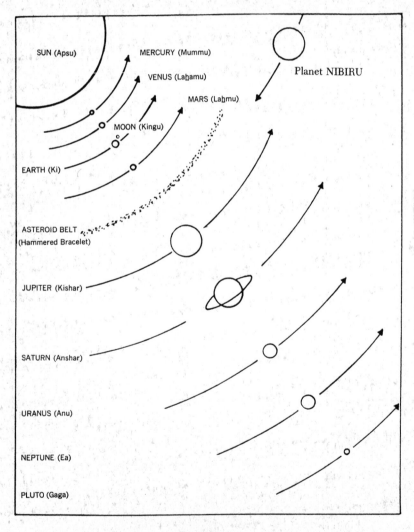

FIGURE 43

with the "artful fashioning" of "Man," with devising the calendar, and with the selection of Babylon to be the "Navel of the Earth."

The New Year festival—the most important religious event of the year—began on the first day of the month Nissan, coinciding with the Spring Equinox. Calling it in Babylon the *Akiti* festival, it evolved there into a twelve-day-long celebration from the Sumerian ten-day A.KI.TI ("On Earth Bring Life") festival. It was conducted according to elaborately defined ceremonies and prescribed rituals that reenacted (in Sumer) the tale of Nibiru and the coming of the Anunnaki to Earth, as well as (in Babylon) the life story of Marduk. It included episodes from the Pyramid Wars, when he was sentenced to die in a sealed tomb, and his "resurrection" when he was brought out of it alive; his exile to become the Unseen; and his final victorious Return. Processions, comings and goings, appearances and disappearances, and even passion plays with actors visually and vividly presented Marduk to the people as a suffering god—suffering on Earth but finally victorious by gaining supremacy through a heavenly counterpart. (The New Testament's Jesus story was so similar that scholars and theologians in Europe debated a century ago whether Marduk was the *"Prototype Jesus"*.)

The ceremonies consisted of two parts. The first involved a solitary boat ride by Marduk upon and across the river, to a structure called *Bit Akiti* ("House of Akiti"); the other took place within the city itself. It is evident that the solitary part symbolized Marduk's celestial travel from the home planet's outer location in space to the inner solar system—a journey in a boat upon waters, in conformity with the concept that interplanetary space was a primeval "Watery Deep" to be traversed by "celestial boats" (spacecraft)—a concept represented graphically in Egyptian art, where the celestial gods were depicted as coursing in the skies in "celestial barques" **(Fig. 44)**.

It was upon Marduk's successful return from the outer and lonely Bit Akiti that the public festivities began. Those public and

FIGURE 44

joyous ceremonies started with the greeting of Marduk at the wharf by other gods, and his accompaniment by the king and priests in a Sacred Procession, attended by ever-larger crowds. The descriptions of the procession and its route were so detailed that they guided the archaeologists who excavated ancient Babylon. From the texts inscribed on clay tablets and from the unearthed topography of the city, it emerged that there were seven stations at which the sacred procession made stops for prescribed rituals. The stations bore both Sumerian and Akkadian names and symbolized (in Sumer) the travels of the Anunnaki within the solar system (from Pluto to Earth, the seventh planet), and (in Babylon) the "stations" in Marduk's life story: his divine birth in the "Pure Place"; how his birthright, his entitlement to supremacy, was denied; how he was sentenced to death; how he was buried (alive, in the Great Pyramid); how he was rescued and resurrected; how he was banished and went into exile; and how in the end even the great gods, Anu and Enlil, bowed to destiny and proclaimed him supreme.

The original Sumerian Epic of Creation extended over six tablets (paralleled by the biblical six days of creation). In the Bible, God rested on the seventh day, using it to review His handiwork. The Babylonian revision of the Epic culminated with the addition of a seventh tablet that was entirely devoted to the glorification of Marduk by the granting to him of fifty names—an act that symbolized the assumption by him of the Rank of Fifty that was until then Enlil's (and to which Ninurta had been in line).

Starting with his traditional name MAR.DUK, "son of the Pure Place," the names—alternating between Sumerian and Akkadian—granted him epithets that ranged from "Creator of All" to "Lord who fashioned Heaven and Earth" and other titles relating to the celestial battle with Tiamat and the creation of the Earth and the Moon: "Foremost of all the gods," "Allotter of tasks to the Igigi and the Anunnaki" and their Commander, "The god who maintains life . . . the god who revives the dead," "Lord of all the lands," the god whose decisions and benevolence sustain Mankind, the people he had fashioned, "Bestower of cultivation," who causes rains to enrich the crops, allocates fields, and "heaps abundance" for gods and people alike.

Finally, he was granted the name NIBIRU, "He who shall hold the Crossing of Heaven and Earth":

> The Kakkabu which in the skies is brilliant . . .
> He who the Watery Deep ceaselessly courses—
> Let "Crossing" be his name!
> May he uphold the courses of the stars in heaven,
> May he shepherd the heavenly gods as sheep.

"With the title 'Fifty' the great gods proclaimed him; He whose name is 'Fifty' the gods made supreme," the long text states in conclusion.

When the nightlong reading of the seven tablets was completed—it probably was dawn by then—the priests who conducted the ritual service made the following prescribed pronouncements:

> Let the Fifty Names be kept in mind . . .
> Let the wise and knowing discuss them.
> Let the father recite them to his son,
> Let the ears of shepherds and herdsmen be opened.

Let them rejoice in Marduk, the "Enlil" of the gods,
whose order is firm, whose command is unalterable;
The utterance of his mouth no god can change.

FIGURE 45

When Marduk appeared in sight of the people, he was dressed in magnificent vestments that put to shame the simple wool garments of the olden gods of Sumer & Akkad **(Fig. 45)**.

Although Marduk was an unseen god in Egypt, his veneration and acceptance there took hold rather quickly. A Hymn to Ra-Amon that glorified the god by a variety of names in emulation of the Akkadian Fifty Names called him "Lord of the gods, who behold him in the midst of the horizon"—a celestial god—"who made the entire Earth," as well as a god on Earth "who created mankind and made the beasts, who created the fruit tree, made herbage and gave life to cattle"—a god "for whom the sixth day is celebrated." The snippets of similarities to the Mesopotamian and the biblical creation tales are clear,

According to these expressions of faith, on Earth, in Egypt, Ra/ Marduk was an unseen god because his main abode was elsewhere— one long hymn actually referred to Babylon as the place where the gods are in jubilation for his victory (scholars, though, assume the reference is not to the Mesopotamian Babylon, but to a town by that name in Egypt). In the heavens he was unseen, because "he is far away in heaven," because he went "to the *rear of the horizons . . . to the height of heaven.*" Egypt's reigning symbol—a Winged Disc usually flanked by serpents—is commonly explained as a Sun disc "because Ra was the Sun"; but, in fact, it was the ancient world's ubiquitous symbol of Nibiru **(Fig. 46)**, and it was Nibiru that has become a distant unseen "star."

FIGURE 46

Because Ra/Marduk was physically absent from Egypt, it was in Egypt that his Star Religion was expressed in its clearest form. There, **Aten**, the "Star of Millions of Years" representing Ra/Marduk in his celestial aspect, became *The Unseen* because it was "far away in heaven," because it had gone "to the rear of the horizon."

The transition to Marduk's New Age and new religion was not so smooth in the Enlilite lands. First, southern Mesopotamia and the western lands that were in the path of the poisonous wind had to recover from its impact.

The calamity that befell Sumer, it will be recalled, was not the nuclear explosion per se but the ensuing radioactive wind. The cities were emptied of their residents and livestock, but were physically undamaged. The waters were poisoned, but the flowing two great rivers soon corrected that. The soil absorbed the radioactive poison, and that took longer to recover; but that, too, improved with time. And so it was possible for people to slowly repopulate and reinhabit the desolated land.

The first recorded administrative ruler in the devastated south was an ex-governor of Mari, a city way northwest on the Euphrates River. We learn that "he was not of Sumerian seed"; his name, Ishbi-Erra, was in fact a Semitic name. He established his headquarters in the city of Isin, and from there he oversaw the efforts to resurrect the other major cities, but the process was slow, difficult, and at times chaotic. His efforts at rehabilitation were continued by several successors, also bearing Semitic names, the so-called "Dynasty of Isin." All together, it took them close to a century to revive Ur, Sumer's economic center, and ultimately Nippur, the land's traditional religious heart; but by then that city-at-a-time process ran into challenges from other local city rulers, and the erstwhile Sumer remained fragmented and a broken land.

Even Babylon itself, though outside the Evil Wind's direct path, needed a revived and repopulated country if it was to rise to imperial size and status, and it did not attain the grandeur of Marduk's prophecies for quite some time. More than a century had to pass until a formal dynasty, called by scholars the First Dynasty of Babylon, was installed on its throne (circa 1900 B.C.E.). Yet another century had to pass until a king who lived up to the prophesied greatness sat on Babylon's throne; his name was Hammurabi. He is mostly known for the code of laws proclaimed by him—laws recorded on a stone stela that archaeologists have discovered (and that is now in the Louvre in Paris).

It still took some two centuries before Marduk's prophetic vision regarding Babylon could come true. The meager evidence from the postcalamity time—some scholars refer to the period following the demise of Ur as a Dark Age in Mesopotamian history—suggests that Marduk let the other gods—even his adversaries—take care of the recovery and repopulation of their own olden cult centers, but it is doubtful that they took up his invitation. The recovery and rebuilding that were started by Ishbi-Erra began at Ur, but there is no mention of Nannar/Sin and Ningal returning to Ur. There is mention of Ninurta's occasional presence in Sumer, especially in regard to its garrisoning by troops from Elam and Gutium, but there is no record that he or his spouse Bau ever returned to their beloved Lagash. The efforts by Ishbi-Erra and his successors to restore the cult centers and their temples culminated—after the passage of seventy-two years—at Nippur, but there is no mention that Enlil and Ninlil resumed residence there.

Where had they gone? One avenue of exploring that intriguing subject was to ascertain what Marduk himself—now supreme and claiming to be the giver of commands to all the Anunnaki—had planned for them.

The textual and other evidence from that time show that

Marduk's rise to supremacy did not end polytheism—the religious beliefs in many gods. On the contrary, his supremacy required continued polytheism, for to be supreme over other gods, the existence of other gods was necessary. He was satisfied to let them be, as long as their prerogatives were subject to his control; a Babylonian tablet recorded (in its undamaged portion) the following list of divine attributes that were henceforth vested in Marduk:

Ninurta	is	Marduk of the hoe
Nergal	is	Marduk of the attack
Zababa	is	Marduk of the combat
Enlil	is	Marduk of lordship and counsel
Sin	is	Marduk the illuminator of the night
Shamash	is	Marduk of justice
Adad	is	Marduk of rains

The other gods remained, their attributes remained—but they now held attributes of Marduk that *he* granted to them. He let their worship be continued; the very name of the interim ruler/administrator in the south, Ishbi-*Erra* ("Priest of Erra," i.e., of Nergal) confirms this tolerant policy. But what Marduk expected was that they come and stay with him in his envisaged Babylon—prisoners in golden cages, one may say.

In his autobiographical *Prophecies* Marduk clearly indicated his intentions in regard to the other gods, including his adversaries: they were to come and reside next to him, in Babylon's sacred precinct. Sanctuaries or pavilions for Sin and Ningal, where they would reside—"together with their treasures and possessions"!—are specifically mentioned. Texts describing Babylon, and archaeological excavations there, show that in accordance with Marduk's wishes, Babylon's sacred precinct also included residence-shrines dedicated to Ninmah, Adad, Shamas, and even Ninurta.

When Babylon finally rose to imperial power—under Hammurabi—its ziggurat-temple indeed reached skyward; the prophesied great king in time did sit on its throne; but to its priest-filled sacred precinct, the other gods did not flock. That manifestation of the New Religion did not come about.

Looking at the Hammurabi stela recording his law code **(Fig. 47)**, we see him receiving the laws from none other than Utu/Shamash—the very one, according to the above-quoted list, whose prerogatives as God of Justice now belonged to Marduk; and the preamble inscribed on the stela invoked Anu *and*

FIGURE 47

Enlil—the one whose "Lordship and Counsel" were presumably taken over by Marduk—as the gods to whom Marduk was beholden for his status:

> *Lofty Anu,*
> *Lord of the gods who from heaven to Earth came,*
> *and Enlil, Lord of Heaven and Earth*
> *who determines the Land's destinies,*

Determined for Marduk, the firstborn of Enki,
the Enlil-functions over all mankind.

These acknowledgments of the continued empowerment of Enlilite gods, two centuries after the Age of Marduk began, reflect the actual state of affairs: They did not come to retire in Marduk's sacred precinct. Dispersed away from Sumer, some accompanied their followers to far lands in the four corners of the Earth; others remained nearby, rallying their followers, old and new, to a renewed challenge to Marduk.

The sense that Sumer as a homeland was no more is clearly expressed in the divine instructions to Abram of Nippur—on the eve of the nuclear upheavaling—to "Semitize" his name to Abraham (and that of his wife Sarai to Sarah), and to make his permanent home in Cannan. Abraham and his wife were not the only Sumerians in need of a new refuge. The nuclear calamity triggered migrational movements on a scale unknown before. The first wave of people was *away* from the affected lands; its most significant aspect, and one with the most lasting effects, was the dispersal of Sumer's remnants away from Sumer. The next wave of migrants was *into* that abandoned land, coming in waves from all directions.

Whichever direction those migration waves took, the fruits of two thousand years of Sumerian civilization were adopted by the other peoples that followed them in the next two millenia. Indeed, though Sumer as a physical entity was crushed, the attainments of its civilization are still with us to this day—just look up your *twelve-month* calendar, check the time on *your watch* that retained the Sumerian sexagesimal ("base sixty") system, or drive in your contraption on *wheels* (a car).

The evidence for a widespread Sumerian diaspora with its language, writing, symbols, customs, celestial knowledge, beliefs, and gods comes in many forms. Beside the generalities—a religion based on a pantheon of gods who have come from the heavens, a divine hierarchy, god epithet-names that mean the same in the different languages, astronomical knowledge that included a home planet of the gods, a zodiac with its twelve houses, virtually identical creation tales, and memories of gods and demigods that scholars treat as "myths"—there are a host of astounding specific similarities that cannot be explained other than by an actual presence of Sumerians. It was expressed in the spread in Europe of Ninurta's Double-Eagle symbol (**Fig. 48**); the fact that three European languages—Hungarian, Finnish, and Basque—are akin only to Sumerian; and the widespread depiction throughout the world—even in South America—of Gilgamesh fighting off with bare hands two ferocious lions (**Fig. 49**).

FIGURE 48

In the Far East, there is the clear similarity between the Sumerian cuneiform writing and the scripts of China, Korea, and Japan. The similarity is not only in the script: many similar glyphs are identically pronounced and also have the same meanings. In Japan, civilization has beeen attributed to an enigmatic forefather-tribe called AINU. The emperor's family has been deemed to be a line of demigods descended from the Sun-god, and the investiture ceremonies of a new king include a secret solitary nightly stay with the Sun goddess—a ritual ceremony

FIGURE 49

that uncannily emulates the Sacred Marriage rites in ancient Sumer, when the new king spent a night with Inanna/Ishtar.

In the erstwhile Four Regions, the migratory waves of diverse peoples triggered by the nuclear calamity and Marduk's New Age, much like flowing and overflowing rivers and rivulets after stormy rains, filled the pages of the ensuing centuries with the rise and fall of nations, states, and city-states. Into the Sumerian void, newcomers came in from near and far; their arena, their central stage, remained what can rightly be called the Lands of the Bible. Indeed, until the advent of modern archaeology, little or nothing was known about most of them except for their mention in the Hebrew Bible; it provided not only a record of those various peoples, but also of their "national gods"—and of the wars fought in the name of those gods.

But then nations such as the Hittites, states such as Mitanni, or royal capitals such as Mari, Carchemish, or Susa, which were doubt-filled puzzles, were literally dug up by archaeology; in their ruins there were found not only telltale artifacts but also thousands of inscribed clay tablets that brought to light both their existence as well as the extent of their debt to the Sumerian legacy. Virtually everywhere, Sumerian "firsts" in sciences and technology, literature and art, kingship and priesthood were the foundation on which subsequent cultures were developed. In astronomy, Sumerian terminology, orbital formulas, planetary lists, and zodiacal concepts were retained. The Sumerian cuneiform script was kept in use for another thousand years, and then more. The Sumerian language was studied, Sumerian lexicons were compiled, and Sumerian epic tales of gods and heroes were copied and translated. And once those nations' diverse languages were deciphered, it turned out that their gods were, after all, members of the old Anunnaki pantheon.

Did the Enlilite gods themselves accompany their followers when such replanting of Sumerian knowledge and beliefs took place in faraway lands? The data are inconclusive. But what is historically certain is that within two or three centuries of the New Age, in lands bordering Babylonia, those who were supposed to become Marduk's retired guests embarked on an even newer kind of religious affiliations: *National State Religions.*

Marduk may have garnered the Fifty divine names; but it did not prevent, from then on, nation fighting nation and men killing men "in the name of God"—*their* god.

Chapter VIII

IN THE NAME OF GOD

If the prophecies and messianic expectations attendant on the New Age of the twenty-first century B.C.E. look familiar to us today, the battle cries of the subsequent centuries would not sound strange, either. If in the third millennium B.C.E. god fought god using armies of men, in the second millennium B.C.E. men fought men "in the name of god."

It took just a few centuries after the start of Marduk's New Age to show that the fulfillment of his prophecies of grandeur would not easily come. Significantly, the resistance came not so much from the dispersed Enlilite gods but from the people, the masses of their loyal worshippers!

More than a century had to pass from the time of the nuclear ordeal until Babylon (the city) emerged on the stage of history as Babylonia (the state) under its First Dynasty. During that interval southern Mesopotamia—the Sumer of old—was left to recover in the hands of temporary rulers headquartered in Isin and then in Larsa; their theophoric names—Lipit-*Ishtar*, Ur-*Ninurta*, Rim-*Sin*, *Enlil*-Bani—flaunted their Enlilite loyalties. Their crowning achievement was the restoration of Nippur's temple exactly seventy-two years after the nuclear havoc—another indication of where their loyalties lay, and of an adherence to a zodiacal time count.

Those non-Babylonian rulers were scions of Semitic-speaking royals from a city-state called Mari. As one looks at a map showing the nation-states of the first half of the second millennium B.C.E. **(Fig. 50)**, it becomes clear that the non-Mardukite states formed a formidable vise around Greater Babylon, starting with Elam and Gutium on the southeast and east; Assyria and Hatti in the north; and as a western anchor in the chain, Mari on the mid-Euphrates.

FIGURE 50

Of them, Mari was the most "Sumerian," even having served once as Sumer's capital, the tenth as that function rotated among Sumer's major cities. An ancient port city on the Euphrates River,

it was a major crossing point for people, goods, and culture between Mesopotamia in the east, the Mediterranean lands in the
west, and Anatolia in the northwest. Its monuments bore the fin-
est examples of Sumerian writing, and its huge central palace was
decorated with murals, astounding in their artistry, honoring
Ishtar **(Fig. 51)**. (A chapter on Mari and my visit to its ruins can
be read in *The Earth Chronicles Expeditions*.)

FIGURE 51

Its royal archive of thousands of clay tablets revealed how
Mari's wealth and international connections to many other
city-states were first used and then betrayed by the emerging
Babylon. After at first attaining the restoration of southern
Mesopotamia by the Mari royals, Babylon's kings—feigning
peace and unprovoked—treated Mari as an enemy. In **1760 B.C.E.**
the Babylonian king Hammurabi attacked, sacked, and destroyed
Mari, its temples and its palaces. It was done, Hammurabi
boasted in his annals, "through the mighty power of Marduk."

After the fall of Mari, chieftains from the "Sealands"—Sumer's
marshy areas bordering the Lower Sea (Persian Gulf)—conducted raids northward, and took from time to time control of
the sacred city of Nippur. But those were temporary gains, and
Hammurabi was certain that his vanquishing of Mari completed
Babylon's political and religious domination of the old Sumer &
Akkad. The dynasty to which he belonged, named by scholars

the First Dynasty of Babylon, began a century before him and continued through his descendants for another two centuries. In those turbulent times, it was quite an achievement.

Historians and theologians agreee that in **1760 B.C.E.** Hammurabi, calling himself "King of the Four Quarters," "put Babylon on the world map" and **launched Marduk's distinct Star Religion.**

When Babylon's political and military supremacy was thus established, it was time to assert and aggrandize its religious domination. In a city whose splendor was extolled in the Bible and whose gardens were deemed one of the ancient world's wonders, the sacred precinct, with the Esagil ziggurat-temple at its center, was protected by its own walls and guarded gates; inside, processional ways were laid out to fit the religious ceremonies, and shrines were built for other gods (whom Marduk expected to be his unwilling guests). When archaeologists excavated Babylon, they found not only the city's remains but also "architectural tablets" describing and mapping out the city; though many of the structures are remains from later times, this artist's conception of the sacred precinct's center **(Fig. 52)** gives a good idea of Marduk's magnificent headquarters.

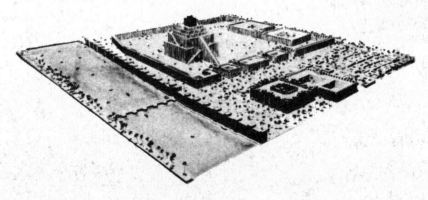

FIGURE 52

As befits a "Vatican," the sacred precinct was also filled with an impressive array of priests whose religious, ceremonial, administrative, political, and menial tasks can be gleaned from their varied groupings, classifications, and designations.

At the bottom of the hierarchy were the service personnel, the *Abalu*—"Porters"—who clean-swept the temple and adjoining buildings, provided the tools and utensils that the other priests required, and acted as general supply and warehousing personnel—except for woollen yarns, which were entrusted only to the *Shu'uru* priests. Special priests, like the *Mushshipu* and *Mulillu*, performed ritual purification services, except that it required a *Mushlahhu* to handle snake infestations. The *Umannu*, Master Craftsmen, worked in workshops where artful religious objects were fashioned; the *Zabbu* were a group of female priestesses, chefs, and cooks who prepared the meals. Other priestesses acted as professional bewailers in funerals; the *Bakate* knew how to shed bitter tears. And then there were the *Shangu*—simply "the priests"—who oversaw the overall functioning of the temple, the smooth performance of its rituals, and the receiving and handling of the offerings, or who were responsible for the gods' clothes; and so on and on.

The provision of personal "butlering" services to the resident gods was handled by a small, specially selected elite group of priests. There were the *Ramaqu* who handled the purification-by-water rituals (honored with bathing the god), and the *Nisaku* who poured out the used water. The anointing of the god with "Sacred Oil"—a delicate mixture of specific aromatic oils—was placed in specialized hands, starting with the *Abaraku* who mixed the ointments, and included the *Pashishu* who performed the anointing (in the case of a goddess the priests were all eunuchs). Then there were altogether other priests and priestesses, including the Sacred Choir—the *Naru* who sang, the *Lallaru* who were singers and musicians, and the *Munabu* whose specialty was lamentations. In each group there was the *Rabu*—the Chief, the one in charge.

As envisaged by Marduk, once his Esagil ziggurat-temple was

raised heavenward, its main function was to constantly observe the heavens; and indeed, the most important segment of temple priests were those whose task it was to observe the heavens, track the movement of stars and planets, record special phenomena (such as a planetary conjunction or an eclipse), and consider whether the heavens bespoke omens; and if so, to interpret what they did portend.

The astronomer-priests, generally called *Mashmashu*, included diverse specialties; a *Kalu* priest, for example, specialized in watching the Constellation of the Bull. It was the duty of the *Lagaru* to keep a detailed daily record of the celestial observations, and to convey the information to a cadre of interpreter-priests. These—making up the top priestly hierarchy—included the *Ashippu*, Omen specialists, the *Mahhu* "who could read the signs," and the *Baru*—"Truth-tellers"—who "understood mysteries and divine signs." A special priest, the *Zaqiqu*, was charged with conveying the divine words to the king. Then at the head of those astronomer-astrologer priests was the *Urigallu*, the Great Priest, who was a holy man, a magician, and a physician, whose white vestments were elaborately color-trimmed at the hems.

The discovery of some seventy tablets that formed a continuous series of observations and their meaning, named after the opening words *Enuma Anu Enlil*, revealed both the transition from Sumerian astronomy and the existence of oracular formulas that dictated what a phenomenon meant. In time a host of diviners, dream interpreters, fortune-tellers, and the like joined the hierarchy, but they were in the king's rather than the gods' service. In time the celestial observations degraded to astrological omens for king and country—predicting war, tranquility, overthrows, long life or death, abundance or pestilences, divine blessings or godly wrath. But in the beginning the celestial observations were purely astronomical and were of prime interest to the god—Marduk—and only derivatively to king and people.

It was not by chance that a Kalu priest specialized in watching Enlil's Constellation of the Bull for any untoward phenomena, for the main purpose of the Esagil-as-observatory was to track the heavens zodiacally and keep an eye on Celestial Time. The fact that significant events prior to the nuclear blast happened in seventy-two-year intervals, and continued to do so afterward (see above and earlier chapters), suggests that the zodiacal clock, in which it took seventy-two years for a Precessional shift of one degree, continued to be observed and adhered to.

It is clear from all the astronomical (and astrological) texts from Babylon that its astronomer-priests retained the Sumerian division of the heavens into three Ways or paths, each occupying sixty degrees of the celestial arc: the Way of Enlil for the northern skies, the Way of Ea for the southern skies, and the Way of Anu as the central band **(Fig. 53)**. It was in the latter that the zodiacal constellations were located, and it was there that "Earth met Heaven"—at the horizon.

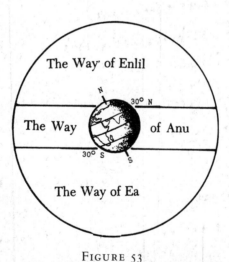

FIGURE 53

Perhaps because Marduk attained supremacy in accordance with Celestial Time, the zodiacal clock, his astronomer-priests continuously scanned the skies at the horizon, the Sumerian AN.UR, "Heaven's Base." There was no point in looking up to the Sumerian AN.PA, "Heaven's Top," the zenith, for Marduk as a "star," Nibiru, was by then gone and unseen.

But as an orbiting planet, though unseen now, it was bound to return. Expressing its equivalent of the Marduk-is-Nibiru theme, the Egyptian version of Marduk's Star-Religion openly promised its faithful that a time will come when this god-star or star-god would *reappear* as the ATEN.

It was this aspect of Marduk's Star Religion—the eventual Return—that directly challenged Babylon's Enlilite adversaries, and shifted the conflict's focus to renewed messianic expectations.

Of the post-Sumer actors on the stage of the Old World, four that grew to imperial status left the deepest imprint on history: Egypt and Babylonia, Assyria and Hatti (the land of the Hittites); and each one had its "national god."

The first two belonged to the Enki-Marduk-Nabu camp; the other two were beholden to Enlil, Ninurta, and Adad. Their national gods were called Ra-Amon and Bel/Marduk, Ashur and Teshub, and it was in the name of those gods that constant, prolonged, and cruel wars were fought. The wars, historians may explain, were caused by the usual reasons for war: resources, territory, need, or greed; but the royal annals that detailed the wars and military expeditions presented them as *religious wars* in which one's god was glorified and the opposite deity humiliated. However, the looming expectations of the Return turned those wars to *territorial campaigns* that had *specific sites as their targets.*

The wars, according to the royal annals of all those lands, were launched by the king "on the command of my god" so-and-so; the campaign was carried out "in accordance with an oracle" from this or that god; and as often as not, victory was attained with the help of unopposable weapons or other direct help provided by the god. An Egyptian king wrote in his war records that it was "Ra who loves me, Amon who favors me," who instructed him to march "against these enemies whom Ra abominates." An Assyrian king, recording the defeat of an enemy king, boasted that he replaced in the city's temple the images of the city's gods "with the images of my gods, and declared them to be henceforth the gods of the country."

A clear example of the religious aspect of those wars—and the deliberate choice of targets—can be found in the Hebrew Bible, in 2 Kings Chapters 18–19, in which the siege of Jerusalem by the army of the Assyrian king Sennacherib is described. Having surrounded and cut off the city, the Assyrian commander engaged in psychological warfare in order to get the city's defenders to surrender. Speaking in Hebrew so that all on the city's walls could understand, he shouted to them the words of the king of Assyria: Don't be deceived by your leaders that your god Yahweh will protect you; "Has any of the gods of the nations ever rescued their lands from the hand of the king of Ashur? Where are the gods of Hamath and Arpad? Where are the gods of Sepharvaim, Hena and Avva? Where are the gods of the land of Samaria? Which of the gods of all these lands ever rescued his land from my hand? Will then Yahweh rescue Jerusalem from my hand?" (Yahweh, the historical records show, did.)

What were those religious wars about? The wars, and the national gods in whose name they were fought, don't make sense except when one realizes that at the core of the conflicts was what the Sumerian had called DUR.AN.KI—the "Bond Heaven-Earth." Repeatedly, the ancient texts spoke of the catastrophe "when Earth was separated from Heaven"—when the spaceport connecting them was destroyed. The overwhelming question in the aftermath of the nuclear calamity was this: **Who—which god and his nation—could claim to be the one on Earth who now possessed the link to the Heavens?**

For the gods, the destruction of the spaceport in the Sinai peninsula was a material loss of a facility that required replacement. But can one imagine the impact—the spiritual and religious impact—on Mankind? *All of a sudden, the worshipped gods of Heaven and Earth were cut off from Heaven . . .*

With the spaceport in the Sinai now obliterated, only three space-related sites remained in the Old World: the Landing Place in the cedar mountains; the post-Diluvial Mission Control Center that replaced Nippur; and the Great Pyramids in Egypt that anchored the Landing Corridor. With the destruction of the spaceport, did those other sites still have a useful celestial function—and thus also a religious significance?

We know the answer, to some extent, because all three sites still stand on Earth, challenging mankind by their mysteries and the gods by facing upward to the heavens.

The most familiar of the three is the Great Pyramid and its companions in Giza **(Fig. 54)**; its size, geometric precision, inner complexity, celestial alignments, and other amazing aspects have long cast doubt on the attribution of its construction to a Pharaoh named Cheops—an attribution supported solely by a discovery of a hieroglyph of his name inside the pyramid. In *The Stairway to Heaven* I offered proof that those markings were a modern forgery, and in that book and others voluminous textual and pictorial evidence was provided to explain how and why the Anunnaki designed and built those pyramids. Having been stripped of its radiating guidance equipment during the wars of the gods, the Great Pyramid and its companions continued to serve as physical beacons for the Landing Corridor. With the spaceport gone, they just remained silent witnesses to a vanished Past; there has been no indication that they ever became sacred religious objects.

The Landing Place in the cedar forest has a different record. Gilgamesh, who went to it almost a millennium before the nuclear calamity, witnessed there the launching of a rocket ship; and the Phoenicians of the nearby city of Byblos on the Mediterranean coast depicted on a coin **(Fig. 55)** a rocket ship emplaced on

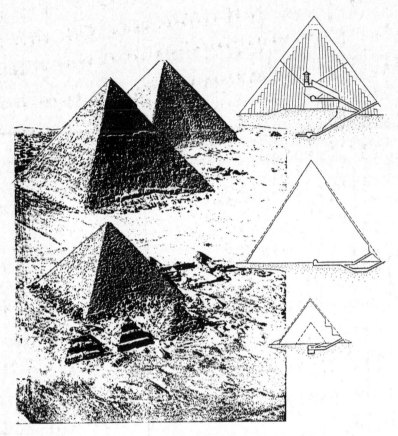

FIGURE 54

FIGURE 55

a special base within an enclosure at the very same place—almost a thousand years after the nuclear event. So, with and then without the spaceport, **the Landing Place continued to be operative.**

The place, *Ba'albek* ("The valley-cleft of Ba'al"), in Lebanon, consisted in antiquity of a vast (about five million square feet) platform of paved stones at the northwestern corner of which an enormous stone structure rose heavenward. Built with perfectly shaped massive stone blocks weighing 600 to 900 tons each, its western wall was especially fortified with the heaviest stone blocks on Earth, including three that weigh an incredible 1,100 tons each and are known as the Trilithon **(Fig. 56)**. The amazing fact about those colossal stone blocks is that they were quarried about two miles away in the valley, where one such block, whose quarrying was not completed, still sticks out from the ground **(Fig. 57)**.

FIGURE 56

FIGURE 57

The Greeks venerated the place since Alexander's time as Heliopolis (City of the Sun god); the Romans built there the greatest temple to Zeus. The Byzantines converted it to a great church; the Moslems after them built there a mosque; and present-day Maronite Christians revere the place as a relic from the Time of the Giants. (A visit to the place and its ruins, and how it functioned as a launch tower, are described in *The Earth Chronicles Expeditions*.)

Most sacred and hallowed to this day has been the site that served as Mission Control Center—*Ur-Shalem* ("City of the Comprehensive God"), **Jerusalem**. There, too, as in Baalbek but on a reduced scale, a large stone platform rests on a rock and cut-stones foundation, including a massive **western wall with three colossal stone blocks** that weigh about six hundred tons *each* **(Fig. 58)**. It was upon that preexisting platform that the Temple to Yahweh was built by King Solomon, its Holy of Holies with the Ark of the

FIGURE 58

Covenant resting upon a sacred rock above a subterranean chamber. The Romans, who built the greatest temple ever to Jupiter in Baalbek, also planned to build one to Jupiter in Jerusalem instead of the one to Yahweh. The Temple Mount is nowadays dominated by the Moslem-built Dome of the Rock **(Fig. 59)**; its gilded dome originally surmonted the Moslem shrine at Baalbek—evidence that the link between the two space-related sites has seldom been missed.

In the trying times after the nuclear calamity, could Marduk's *Bab-Ili*, his "Gateway of the gods," substitute for the olden Bond Heaven-Earth sites? Could Marduk's new Star Religion offer an answer to the perplexed masses?

The ancient search for an answer, it seems, has continued to our very own time.

The most unremitting adversary of Babylon was the Assyrians. Their province, in the upper region of the Tigris River, was called Subartu in Sumerian times and was the northernmost extension of Sumer & Akkad. In language and racial origins they appear to have

FIGURE 59

had a kinship to Sargon of Akkad, so much so that when Assyria became a kingdom and imperial power, some of its most famous kings took the name *Sharru-kin*—Sargon—as their royal name.

All that, gleaned from archaeolgical finds in the past two centuries, corroborates the succint statements in the Bible (Genesis, Chapter 10) that listed the Assyrians among the descendants of Shem, and Assyria's capital Nineveh and other principal cities as "coming out of"—an outgrowth, an extension of—Shine'ar (Sumer). Their pantheon was the Sumerian pantheon—their gods were the Anunnaki of Sumer & Akkad; and the theophoric names of Assyrian kings and high officials indicated reverence to the gods Ashur, Enlil, Ninurta, Sin, Adad, and Shamash. There were temples to them, as well as to the goddess Inanna/Ishtar, who was also extensively worshipped; one of her best-known depictions, as a helmeted pilot **(Fig. 60)**, was found in her temple in Ashur (the city).

Historical documents from the time indicate that it was the Assyrians from the north who were the first to challenge Mar-

FIGURE 60

duk's Babylon militarily. The very first recorded Assyrian king, Ilushuma, led circa 1900 B.C.E. a successful military expedition down the Tigris River all the way south to the border of Elam. His inscriptions state that his aim was to "set the freedom of Ur and Nippur"; and he did remove, for a while, those cities from Marduk's grip.

That was only the first fight between Assyria and Babylonia in a conflict that continued for more than a thousand years and lasted to the end of both. It was a conflict in which the Assyrian kings were usually the aggressors. Neighboring each other, speaking the same Akkadian language, and both inheriting the Sumerian foundation, the Assyrians and Babylonians were distinguishable by just one key difference: their national god.

Assyria called itself the "Land of the god Ashur" or simply *ASHUR*, after the name of its national god, for its kings and people considered this religious aspect to be all that mattered. Its first capital was also called "City of Ashur," or simply *Ashur*. The name meant "The One Who Sees" or "The One Who Is Seen."

Yet with all the countless hymns, prayers, and other references to the god Ashur, it remains unclear who exactly, in the Sumerian-Akkadian pantheon, he was. In god lists he was the equivalent of Enlil; other references sometimes suggest that he was Ninurta, Enlil's son and heir; but since whenever the spouse was listed or mentioned she was always called Ninlil, the conclusion tends to be that the Assyrian "Ashur" was Enlil.

The historical record of Assyria is one of conquest and aggression against many other nations and their gods. Their countless military campaigns ranged far and wide, and were carried on, of course, "in the name of god"—their god, Ashur: "On the command of my god Ashur, the great lord" was the usual opening statement in the Assyrian kings' record of a military campaign. But when it came to the warfare with Babylon, the amazing aspect of Assyria's attacks was its central aim: **not just the rollback of Babylon's influence—but the actual, physical removal of Marduk himself from his temple in Babylon!**

The feat of capturing Babylon and taking Marduk into captivity was first achieved, however, not by the Assyrians but by their neighbors to the north—the Hittites.

Circa 1900 B.C.E. the Hittites began to spread out from their strongholds in north-central Anatolia (today's Turkey), became a major military power, and joined the chain of Enlilite nation-states opposed to Marduk's Babylon. In a relatively short time, they attained imperial status and their domains extended southward to include most of the biblical Canaan.

The archaeological discovery of the Hittites, their cities, records, language, and history, is an astounding and exciting tale of bringing to life and corroborating the existence of people and places hitherto known only from the Hebrew Bible. Hittites are repeatedly mentioned in the Bible, but without the disdain or scorn reserved for worshippers of pagan gods. It refers to their presence throughout the lands where the story and history of the Hebrew Patriarchs unfolded. They were Abraham's neighbors in Harran, and it was from

Hittite landowners in Hebron, south of Jerusalem, that he bought the Machpelah burial cave. Bathsheba, whom King David coveted in Jerusalem, was the wife of a Hittite captain in his army; and it was from Hittite farmers (who used the site for wheat thrashing) that David bought the platform for the Temple on Mount Moriah. King Solomon bought chariot horses from Hittite princes, and it was one of their daughters whom he married.

The Bible considered the Hittites to belong, genealogically and historically, to the peoples of Western Asia; modern scholars believe that they were migrants to Asia Minor from elsewhere—probably from beyond the Caucasus mountains. Because their language, once deciphered, was found to belong to the Indo-European group (as do Greek on the one hand and Sanskrit on the other hand), they are considered to have been non-Semitic "Indo-Europeans." Yet, once settled, they added the Sumerian cuneiform script to their own distinct script, included Sumerian "loan words" in their terminology, studied and copied Sumerian "myths" and epic tales, and adopted the Sumerian pantheon—including the count of twelve "Olympians." In fact, some of the earliest tales of the gods on Nibiru and coming from Nibiru were discovered only in their Hittite versions. The Hittite gods were undoubtedly the Sumerian gods, and monuments and royal seals invariably showed them accompanied by the ubiquitous symbol of the Winged Disc (see Fig. 46), the symbol for Nibiru. These gods were sometimes called in the Hittite texts by their Sumerian or Akkadian names—we find Anu, Enlil, Ea, Ninurta, Inanna/Ishtar, and Utu/Shamash repeatedly mentioned. In other instances the gods were called by Hittite names; leading them was the Hittite national god, **Teshub**—"the Windblower" or "God of storms." He was none other than Enlil's youngest son ISHKUR/Adad. His depictions showed him holding the lightning bolt as his weapon, usually standing upon a bull—the symbol of his father's celestial constellation **(Fig. 61)**.

The biblical references to the extended reach and military prowess of the Hittites were confirmed by archaeological discoveries

FIGURE 61

both at Hittite sites and in the records of other nations. Significantly, the Hittite southward reach embraced the two space-related sites of the Landing Place (today's Baalbek) and the post-Diluvial Mission Control Center (Jerusalem); it also brought the Enlilite Hittites to within striking distance of Egypt, the land of Ra/ Marduk. The two sides thus had all it took to engage in armed conflict. In fact, the wars between the two included some of the ancient world's most famous battles fought "in the name of god."

But rather than attack Egypt, the Hittites sprung a surprise. The first, perhaps, to introduce horse-driven chariots in military campaigns, the Hittite army, totally unexpectedly, in 1595 B.C.E., swept down the Euphrates River, captured Babylon, **and took Marduk into captivity.**

Though one wishes that more detailed records from that time and event would have been discovered, what is known indicates that the Hittite attackers did not intend to take over and rule Babylon: they retreated soon after they had breached the city's defenses and entered its sacred precinct, taking Marduk with them, leaving him unharmed, but apparently under guard, in a city called Hana—a place (yet to be excavated) in the district of Terka, along the Euphrates River.

The humiliating absence of Marduk from Babylon lasted twenty-four years—exactly the same time that Marduk had been in exile in Harran five centuries earlier. After several years of confusion and disorder, kings belonging to a dynasty called the Kassite Dynasty took control of Babylon, restored Marduk's shrine, "took the hand of Marduk," and returned him to Babylon. Still, the Hittite sack of Babylon is considered by historians to have marked the end both of the glorious First Dynasty of Babylon and of the Old Babylonian Period.

The sudden Hittite thrust to Babylon and the temporary removal of Marduk remain an unresolved historical, political, and religious mystery. Was the intention of the raid just to embarrass and diminish Marduk—deflate his ego, confuse his followers—or was there a more far-reaching purpose—or cause—behind it?

Was it possible that Marduk fell victim to the proverbial "hoist by his own petard"?

Chapter IX

THE PROMISED LAND

The capture and removal of Marduk from Babylon had geopolitical repercussions, shifting for several centuries the center of gravity from Mesopotamia westward, to the lands along the Mediterranean Sea. In religious terms, it was the equal of a tectonic earthquake: in one blow, all the great expectations by Marduk for all gods to be gathered under his aegis, and all the messianic expectations by his followers, were gone like a puff of smoke.

But both geopolitically and religiously, the greatest impact can be summed up as the story of three mountains—the three space-related sites that put the Promised Land in the midst of it all: Mount Sinai, Mount Moriah, and Mount Lebanon.

Of all the events that followed the unprecedented occurrence in Babylon, the central and most lasting one was the Israelite Exodus from Egypt—when, for the first time, sites that until then were the gods' alone were entrusted to people.

When the Hittites who took Marduk captive withdrew from Babylon, they left behind political disarray and a religious enigma: How could that happen? Why did it happen? When bad things happened to people, they would say that the gods were angry; so what now that bad things happened to gods—to Marduk? *Was there a God supreme to the supreme god?*

In Babylon itself, the eventual release and return of Marduk did not provide an answer; in fact, it increased the mystery, for the "Kassites" who welcomed the captured god back to Babylon were non-Babylonian strangers. They called Babylon "Karduniash" and had names such as Barnaburiash and Karaindash, but little else is known about them or their original language. To this day it is not clear from where they came and why their kings were allowed to replace the Hammurabi dynasty circa **1660 B.C.E.** and to dominate Babylon from **1560 B.C.E.** until **1160 B.C.E.**

Modern scholars speak of the period that followed Marduk's humiliation as a "dark age" in Babylonian history, not only because of the disarray it caused but mainly because of the paucity of written Babylonian records from that time. The Kassites quickly integrated themselves into the Sumerian-Akkadian culture, including language and cuneiform script, but were neither the meticulous recordkeepers the Sumerians had been nor the likes of previous Babylonian writers of royal annals. Indeed, most of the few royal records of Kassite kings have been found not in Babylon but in Egypt—clay tablets in the El-Amarna archive of royal correspondence. Remarkably, in those tablets the Kassite kings called the Egyptian Pharaohs "my brother."

The expression, though figurative, was not unjustified, for Egypt shared with Babylon the veneration of Ra-Marduk and, like Babylonia, had also undergone a "dark age"—a period scholars call the Second Intermediate Period. It began with the demise of the Middle Kingdom circa 1780 B.C.E. and lasted until about **1560 B.C.E.** As in Babylonia, it featured a reign of foreigner kings known as "Hyksos." Here, too, it is not certain who they were, from where they came, or how it was that their dynasties were able to rule Egypt for more than two centuries.

That the dates of this Second Intermediate Period (with its many obscure aspects) parallel the dates of Babylon's slide from the peak of Hammurabi's victories (**1760** B.C.E.) to the capture and resumption of Marduk's worship in Babylon (circa **1560** B.C.E.) is probably neither accident nor coincidence: those similar developments at parallel times in Marduk's principal lands happened because Marduk was "hoist by his own petard"—the very justification for his claim to supremacy was now causing his undoing.

The "petard" was Marduk's own initial contention that the time for his supremacy on Earth had arrived because in the heavens the Age of the Ram, his age, had arrived. But as the zodiacal clock kept ticking, the Age of the Ram started to slowly slip away. The physical evidence from those perplexing times still exists, and can be seen, in Thebes, the ancient Egyptian capital of Upper Egypt.

Apart from the great pyramids of Giza, ancient Egypt's most impressive and majestic monuments are the colossal temples of Karnak and Luxor in southern (Upper) Egypt. The Greeks called the place Thebai, from which its name in English—Thebes—derives; the ancient Egyptians called it the City of Amon, for it was to this

FIGURE 62

unseen god that those temples were dedicated. The hieroglyphic writing and the pictorial depictions on their walls, obelisks, pylons, and columns **(Fig. 62)** glorify the god and praise the Pharaohs who built, enlarged, expanded—and kept changing—the temples. It was there that the arrival of the Age of the Ram was announced by the rows of ram-headed sphinxes (see Fig. 39); and it is there that the very layout of the temples reveals the secret quandary of Egypt's followers of Ra-Amon/Marduk.

One time, visiting the sites with a group of fans, I stood in the midst of a temple waving my hands as a traffic policeman; amazed onlookers wondered, "Who is this nut?," but I was trying to point out to my group the fact that the Thebes temples, built by a succession of Pharaohs, kept changing their orientation **(Fig. 63)**. It was Sir Norman Lockyer who, in the 1890s, first grasped the significance of this architectural aspect, giving rise to a discipline called Archaeoastronomy.

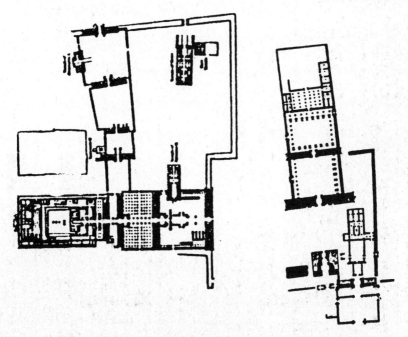

FIGURE 63

Temples that were oriented to the equinoxes, like Solomon's temple in Jerusalem, **(Fig. 64)** (and the old St. Peter's basilica at the Vatican in Rome), faced permanently east, welcoming sunrise on equinox day year after year without reorientation. But temples oriented to the solstices, like Egypt's temples in Thebes or China's Temple of Heaven in Beijing, needed periodic reorientation because due to Precession, where the Sun rises on

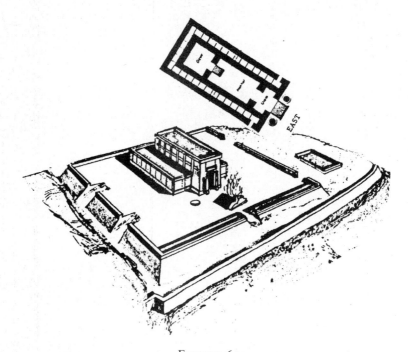

FIGURE 64

solstice day shifts ever so lightly over the centuries—as can be illustrated by Stonehenge, where Lockyer applied his findings (see Fig. 6). The very temples that Ra/Marduk's followers had erected to glorify him were showing that the heavens were uncertain about the durability of the god and his Age.

Marduk himself—so aware of the zodiacal clock when he had claimed in the previous millennium that his time had arrived—tried to shift the religious focus by introducing the Star Religion of "Marduk is Nibiru." But his capture and humiliation now raised questions regarding this unseen celestial god. The question, Until when will the Age of Marduk last? changed to the question: If celestially Marduk is the unseen Nibiru, when will it reveal itself, reappear, *return*?

As unfolding events showed, both the religious and the geopolitical focus shifted in the middle of the second millenium B.C.E. to the stretch of land that the Bible called **Canaan**. As the *return of Nibiru* started to emerge as the religious focus, the *space-related sites* also emerged into sharper focus, and it was in the geographic "Canaan" where both the Landing Place and the erstwhile Mission Control Center were located.

Historians tell the ensuing events in terms of the rise and fall of nation-states and the clash of empires. It was circa **1460 B.C.E.** that the forgotten kingdoms of Elam and Anshan (later known as Persia, east and southeast of Babylonia) joined to form a new and powerful state, with Susa (the biblical Shushan) as the national capital and Ninurta, the national god, as *Shar Ilani*—"Lord of the gods"; that newly assertive nation-state was to play a decisive role in ending Babylon's and Marduk's supremacy.

It was probably no coincidence that at about the same time, a new powerful state arose in the Euphrates region where Mari had once dominated. There the biblical Horites (scholars call them Hurrians) formed a powerful state named **Mitanni**—"The Weapon of Anu"—which captured the lands that are now Syria and Lebanon and posed a geopolitical and religious challenge to Egypt. That challenge was countered, most ferociously, by Egypt's Pharaoh Tothmosis III, whom historians describe as an "Egyptian Napoleon."

Interwined with all that was the *Israelite exodus from Egypt*, that period's seminal event, if for no other reason than due to its lasting effects, to this day, on Mankind's religions, social and moral codes, and the centrality of Jerusalem. Its timing was not accidental, for all those developments related to the issue of **who shall control the space-related sites when Nibiru's return will occur.**

As was shown in previous chapters, Abraham did not just happen to become a Hebrew Patriarch, but was a chosen participant in major international affairs; and the places where his tale took us—Ur, Harran, Egypt, Canaan, Jerusalem, the Sinai, Sodom and Gomorrah—were principal sites of the universal story of gods and men in earlier times. The Israelite Exodus from Egypt, recalled and celebrated by the Jewish people during the Passover holiday, was likewise an integral aspect of the events that were then unfolding throughout the ancient lands. The Bible itself, far from treating the Exodus as just an "Israelite" story, clearly placed it in the context of Egyptian history and the international events of the time.

The Hebrew Bible opens the story of the Israelite exodus from Egypt in its second book, *Exodus,* by reminding the reader that the Israelite presence in Egypt began when Jacob (who was renamed *Israel* by an angel) and his other eleven sons joined Jacob's son Joseph in Egypt, in 1833 B.C.E. The full story of how Joseph, separated from his family, rose from being a slave to the rank of viceroy, and how he saved Egypt from a devastating famine, is told by the Bible in the last chapters of *Genesis;* and my take on how Joseph saved Egypt and what evidence of that exists to this day is told in *The Earth Chronicles Expeditions.*

Having reminded the reader of how and when the Israelite presence in Egypt began, the Bible makes it clear that all that was gone and forgotten by the time of the Exodus: "Joseph and all his brothers and all that generation had passed away." Not only they but even the dynasty of the Egyptian kings who were connected to those times were also long gone. A new dynasty came into power: "And there arose a new king over Egypt who knew not Joseph."

Accurately, the Bible describes the change of government in Egypt. The dynasties of the Middle Kingdom based in Memphis were gone, and after the disarray of the Second Intermediate Period the princes of Thebes launched the dynasties of the New Kingdom. Indeed, there arose entirely new kings over Egypt— new dynasties in a new capital, "and they knew not Joseph."

Forgetting the Israelite contribution to Egypt's survival, a new Pharaoh now saw danger in their presence. He ordered a series of oppressive steps against them, including the killing of all male babies. These were his reasons:

> And he said unto his people:
> "Behold, a nation, Children of Israel, is greater and
>   mightier than us;
> Let us deal wisely with them, lest they multiply
> and, when war shall be called, they will join our enemies,
> and fight against us, and leave the land."
> EXODUS 1:9–10

Biblical scholars have assumed all along that the feared nation of the "Children of Israel" were the Israelites sojourning in Egypt. But this is in accord with neither the numbers given nor with the literal wording in the Bible. *Exodus* begins with a list of the names of Jacob and his children who had come, with their children, to join Joseph in Egypt, and states that "all of those who descended from the loins of Jacob, excluding Joseph who was already in Egypt, numbered seventy." (That together with Jacob and Joseph the number totaled 72 is an intriguing detail to ponder.) The "sojourn" lasted four centuries, and according to the Bible the number of all the Israelites leaving Egypt was 600,000; no Pharaoh would consider such a group "greater and mightier than us." (For the identity of that Pharaoh and of "the Pharaoh's Daughter" who raised Moses as her son, see *Divine Encounters*.)

The narrative's wording records the Pharaoh's fear that at time of war, the Israelites will "join our enemies, and fight against us, *and leave the land*." It is a fear not of a "Fifth Column" inside Egypt, but of Egypt's indigent "Children of Israel" leaving to reinforce an enemy nation to whom they are related—all of them being, in Egyptian eyes, "Children of Israel." But what other nation of "Children of Israel" and what war was the Egyptian king talking about?

Thanks to archaeological discoveries of royal records from both sides of those ancient conflicts and the synchronization of their contents, we now know that the New Kingdom Pharaohs were engaged in prolonged warfare against **Mitanni**. Starting circa **1560 B.C.E.** with the Pharaoh Ahmosis, continued by the Pharaohs Amenophis I, Thothmosis I, and Thothmosis II, and intensifying under Thothmosis III through **1460 B.C.E.**, Egyptian armies thrust into Canaan and advanced northward against Mitanni. The Egyptian chronicles of those battles frequently mention *Naharin* as the ultimate target—the Khabur River area, which the Bible called *Aram-Naharayim* ("The Western Land of the Two Rivers"); its principal urban center was Harran!

It was there, Bible students will recall, that Abraham's brother Nahor stayed on when Abraham proceeeded to Canaan; it was from there that Rebecca, the bride of Abraham's son Isaac, came—she was in fact the granddaughter of Nahor. And it was to Harran that Isaac's son Jacob (renamed *Israel*) went to find a bride—ending up marrying his cousins, the two daughters (Le'ah and Rachel) of Laban, the brother of his mother Rebecca.

These direct family ties between the "Children of Israel" (i.e., of Jacob) who were in Egypt and those who stayed on in Naharin-Naharayim are highlighted in the very first verses in *Exodus*: the list of the sons of Jacob who had come to Egypt with him includes the youngest, *Ben-Yamin* (Benjamin), the only full brother of Joseph because both were Jacob's sons by Rachel (the others were sons of Jacob by his wife Le'ah and two concubines). We now know from Mitannian tablets that the most important tribe in the Khabur River area were called *Ben-Yamins!* The name of Joseph's full brother was thus a Mitannian tribal name; no wonder, then, that the Egyptians considered the "Children of Israel" in Egypt and the "Children of Israel" in Mitanni as one combined nation "greater and mightier than us."

That was the war the Egyptians were preoccupied with and that was the reason for the Egyptian military concern—not the small number of Israelites in Egypt if they stayed, but a threat if they "left the land" and occupied territory to the north of Egypt. Indeed, *preventing* the Israelites from leaving appears to have been the central theme of the developing drama of the Exodus—there were the repeated appeals by Moses to the reigning Pharaoh to "let my people go," and the Pharaoh's repeated refusals to grant that request—in spite of ten consecutive divine punishments. Why? **For a plausible answer we need to insert the space connection into the unfolding drama.**

In their northward thrusts, the Egyptians marched through the Sinai peninsula via the Way of the Sea, a route (later called by the Romans *Via Maris*) that afforded passage through the gods' Fourth Region along the Mediterranean coast, without actually entering the peninsula proper. Then, advancing north through Canaan, the Egyptians repeatedly reached the Cedar Mountains of Lebanon and fought battles at *Kadesh,* "The Sacred Place." Those were battles, we suggest, for control of the two sacred space-related sites—the erstwhile Mission Control Center (Jerusalem) in Canaan and the Landing Place in Lebanon. The Pharaoh Thothmosis III, for example, in his war annals, referred to Jerusalem (*"Ia-ur-sa"*), which he garrisoned as the *"place reaching to the outer ends of the Earth"*—a "Navel of the Earth." Describing his campaigns farther north, he recorded battles at Kadesh and Naharin and spoke of taking the Cedar Mountains, the *"Mountains of god's land"* that *"support the pillars to heaven."* The terminology unmistakably identifies by their space-related attributes the two sites he was claiming to have captured "for the great god, my father Ra/Amon."

And the purpose of the Exodus? In the words of the biblical God himelf, to keep His sworn promise to Abraham, Isaac, and Jacob to grant to their descendants as "an Everlasting Heritage" (*Exodus* 6: 4–8); "from the Brook of Egypt to the River Euphrates, the great river"; "the whole of the Land of Canaan," (*Genesis* 15:18, 17: 8); "the Western Mount . . . the Land of Canaan and Lebanon" (*Deuteronomy* 1: 7); "from the desert to Lebanon, from the River Euphrates unto the Western Sea" (*Deuteronomy* 11:24)—even the "*fortified places reaching heavenwards*" wherein "descendants of the *Anakim*"—the Anunnaki—still resided (*Deuteronomy* 9: 1–2).

The promise to Abraham was renewed at the Israelites' first stop, at *Har Ha-Elohim*, the "Mount of the Elohim/gods." And the mission was to take hold, possess, the two other space-related sites, which the Bible repeatedly connected (as in Psalms 48:3), calling Mount Zion in Jerusalem *Har Kodshi*, "My Sacred Mount," and the other, on the crest of Lebanon, *Har Zaphon*, "The Secret North Mount."

The Promised Land clearly embraced both space-related sites; its division among the twelve tribes granted the area of Jerusalem to the tribes of Benjamin and Judah, and the territory that is now Lebanon to the tribe of Asher. In his parting words to the tribes before he died, Moses reminded the tribe of Asher that the northern space-related site was in their land—like no other tribe, he said, they will see the "***Rider of the clouds soaring heavenwards***" (*Deuteronomy* 33: 26). Apart from the territorial assignment, **the words of Moses imply that the site would be functional and used for soaring heavenward in the future.**

Clearly and most emphatically, the Children of Israel were to be the custodians of the two remaining space-related sites of the Anunnaki. That Covenant with the people chosen for the task was renewed, at the greatest theophany on record, at **Mount Sinai.**

It was certainly not by chance that the theophany occurred there. From the very beginning of the Exodus tale—when God called out to Moses and gave him the Exodus assignment—that place in the Sinai peninsula occupied center stage. We read in *Exodus* 3:1 that it happened at the "Mount of the *Elohim*"—the mountain associated with the Anunnaki. The route of the Exodus **(Fig. 65)** was divinely determined, the Israelite multitude having

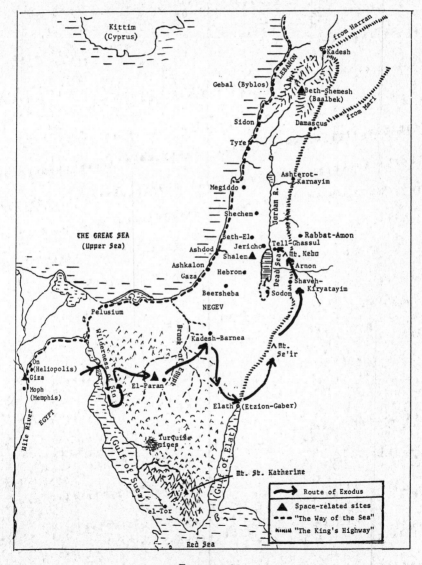

FIGURE 65

been shown the way by a "pillar of cloud by day and a pillar of fire by night." The Children of Israel "journeyed in the wilderness of Sinai according to the instructions of Yahweh," the Bible clearly states; in the third month of the journey they "reached and encamped opposite the Mount"; and on the third day thereafter, Yahweh in his *Kabod* "came down upon Mount Sinai in full view of all the people."

It was the same mount that Gilgamesh, arriving at the place where the rocket ships ascended and descended, had called "Mount *Mashu*." It was the same mount with "the double doors to heaven" to which Egyptian Pharaohs went in their Afterlife Journey to join the gods on the "planet of millions of years." It was the Mount astride the erstwhile Spaceport—and it was there that the Covenant was renewed with the people chosen to be the guardians of the two remaining space-related sites.

As the Israelites were preparing, after the death of Moses, to cross the Jordan River, the boundaries of the Promised Land were restated to the new leader, Joshua. Embracing the locations of the space-related sites, the boundaries emphatically included Lebanon. Speaking to Joshua, the biblical God said:

> *Now arise and cross this Jordan,*
> *thou and all this people, the Children of Israel,*
> *unto the land which I do give to them.*
> *Every place where the soles of your feet shall tread upon*
> *have I given to you, just as I have spoken to Moses:*
> *From the Desert to the Lebanon,*
> *and from the great river, the River Euphrates,*
> *in the country of the Hittites,*
> *unto the Great Sea, where the sun sets—*
> *That shall be your boundary.*
>
> JOSHUA I: 2–4

With so much of the current political, military, and religious turmoil taking place in the Lands of the Bible, and with the Bible itself serving as a key to the past and to the future, one must point out a caveat inserted by the biblical God in regard to the Promised Land. The boundaries, running from the Wilderness in the south to the Lebanon range in the north, and from the Euphrates in the east to the Mediterranean Sea in the west, were reconfirmed to Joshua. These, God said, were the *promised* boundaries. But to become an actual land grant, it had to be obtained by *possession*. Akin to the "planting of the flag" by explorers in the recent past, the Israelites could possess and keep land where they actually set foot—"tread with the soles of their feet"; therefore, God commanded the Israelites not to tarry and delay, but to cross the Jordan and fearlessly and systematically settle the Promised Land.

But when the twelve tribes under the leadership of Joshua were done with the conquest and settlement of Canaan, only part of the areas east of the Jordan were occupied; nor were all of the lands west of the Jordan captured and settled. As far as the two space-related sites were concerned, their stories are totally different: Jerusalem—which was specifically listed (*Joshua* 12: 10, 18: 28)—was firmly in the hands of the tribe of Benjamin. But whether the northward advance attained the Landing Place in Lebanon is in doubt. Subsequent biblical references to the site called it the "Crest of *Zaphon*" (the "secret north place")—what the area's dwellers, the Canaanite-Phoenicians, also called it. (Canaanite epics deemed it to be a sacred place of the god Adad, Enlil's youngest son.)

The crossing of the Jordan River—an accomplishment attained with the help of several miracles—took place "opposite Jericho," and the fortified city of Jericho (west of the Jordan) was the Israelites' first target. The story of the tumbling of its walls and its capture includes a biblical reference to Sumer (*Shin'ar* in Hebrew): in spite of the commandment to take no booty, one of the Israelites could not resist the temptation to "keep a valued garment of Shin'ar."

The capture of Jericho, and the town of Ai south of it, opened the way to the Israelites' most important and immediate target: Jerusalem, where the Mission Control platform had been. The missions of Abraham and his descendants and God's covenants with them never lost sight of that site's centrality. As God told Moses, *it is in Jerualem that His earthly abode was to be*; now the promise-prophecy could be fulfilled.

The capture of the cities on the way to Jerusalem, along with the hill towns surrounding it, turned out to be a formidable challenge, primarily because some of them, and especially Hebron, were inhabited by "children of the *Anakim*"—descendants of the Anunnaki. Jerusalem, it will be recalled, ceased to function as Mission Control Center when the spaceport in the Sinai was wiped out more than six centuries earlier. But according to the Bible, the descendants of the Anunnaki who had been stationed there were still residing in that part of Canaan. And it was "Adoni-Zedek, king of Jerusalem" who formed an alliance with four other city-kings to block the Israelite advance.

The battle that ensued, at Gibe'on in the Valley of Ayalon just north of Jerusalem, took place on a unique day—*the day the Earth stood still*. For the better part of that day, "the Sun stopped and the Moon stood still" (*Joshua* 10: 10–14), enabling the Israelites to win that crucial battle. (A parallel but reverse occurrence, when nighttime lasted an extra twenty hours, took place on the other side of the world, in the Americas; we discuss the matter in *The Lost Realms*.) In the biblical view, then, God himself assured that Jerusalem would come into Israelite hands.

No sooner was kingship established under David than he was commanded by God to clear the platform atop Mount Moriah and sanctify it for Yahweh's Temple. And ever since Solomon built that Temple there, Jerusalem/Mount Moriah/the Temple Mount have remained uniquely sacred. There is, indeed, no other explanation why Jerusalem—a city not at major crossroads, far from waterways, with no natural resources—has been coveted and sacred since antiquity, deemed to be a singular city, a "Navel of the Earth."

The comprehensive list of captured cities given in *Joshua* Chapter 12 names Jerusalem as the third city, following Jericho and Ai, as firmly in Israelite hands. The story was different, however, in regard to the northern space-related site.

The Cedar Mountains of Lebanon run in two ranges, the Lebanon on the west and the anti-Lebanon on the east, separated by the *Bekka*—the "Cleft," a canyon-like valley that has been known since Canaanite times as the "Lord's Cleft" or *Ba'al-Bekka*—hence Ba'albek, the current name of the site of the Landing Place (on the edge of the eastern range, facing the valley). The kings of the "Mount of the North" are listed in the *Book of Joshua* as having been defeated; a place called *Ba'al Gad* "in the valley of Lebanon" is listed as captured; but whether Ba'al-Gad "in the valley of Lebanon" is just another name for Ba'al-Bekka is uncertain. We are told (*Judges* 1: 33) that the Tribe of Naphtali "did not disinherit the dwellers of Beth-Shemesh" ("Abode of Shamash," the Sun god), and that could be a reference to the site, for the later Greeks called the place Heliopolis, "City of the Sun." (Though later the territories under Kings David and Solomon extended to include Beth-Shemesh, it was only temporarily so.)

The primary failure to establish Israelite hegemony over the northern space-related site made it "available" to others. A century and a half after the Exodus the Egyptians attempted to take possession of that "available" Landing Place, but were met by an opposing Hittite army. The epic battle is described in words and illustrations **(Fig. 66)** on the walls of Karnak's temples. Known as the Battle of Kadesh, it ended with an Egyptian defeat, but the war and the battle exhausted both sides so much that the site the Landing Place was left in the hands of the local Phoenician kings of Tyre, Sidon, and Byblos (the biblical Gebal).

FIGURE 66

(The prophets Ezekiel and Amos, who called it "the place of the gods" as well as "the Eden Abode," recognized it as belonging to the Phoenicians.)

The Phoenician kings of the first millennium B.C.E. were well aware of the site's significance and purpose—witness its depiction on a Phoenician coin from Byblos (see Fig. 55). The Prophet Ezekiel (28: 2, 14) admonished the king of Tyre for haughtily believing that, having been to that sacred site of the *Elohim*, he had become himself a god:

> *Thou hast been to a sacred mount,*
> *As a god werest thou, moving within the fiery stones . . .*
> *And you became haughty, saying:*
> *"A god am I, at the place of the* Elohim *I was";*
> *But you are just Man, not god.*

It was at that time that the Prophet Ezekiel—in exile in the "old country," near Harran on the Khabur River—saw divine visions and a celestial chariot, a "Flying Saucer," but that tale must be postponed to a later chapter. Here it is important to note **that of the two space-related sites, only Jerusalem was retained by the followers of Yahweh.**

The first five books of the Hebrew Bible, known as the *Torah* ("The Teachings"), cover the story from Creation, Adam, and Noah to the Patriarchs and Joseph in *Genesis*. The other four books—*Exodus, Leviticus, Numbers,* and *Deuteronomy*—tell the story of the Exodus on the one hand, and on the other hand enumerate the rules and regulations of the new religion of Yahweh. That a new religion encompassing a new, a "priestly" way of life was promulgated is explicitly made clear: "You shall neither do what is done in the land of Egypt, where you had dwelt, nor as is wont in the Land of Canaan whence I am bringing you; you shall neither behave like them nor follow their statutes" (*Leviticus* 18: 2–3).

Having established the basics of the faith ("You shall have no other God before me") and its moral and ethical code in just Ten Commandments, there follow page after page of detailed dietary requirements, rules for priestly rites and vestments, medical teachings, agricultural directives, architectural instructions, family and sexual conduct regulations, property and criminal laws, and so on. They reveal extraordinary knowledge in virtually every scientific discipline, expertise in metals and textiles, acquaintance with legal systems and societal issues, familiarity with the lands, history, customs, and gods of other nations—and certain numerological preferences.

The theme of *twelve*—as in the twelve tribes of Israel or in the twelve-month year—is obvious. Obvious, too, is the predilection for *seven*, most prominently in the realm of festivals and rituals, and in establishing a week of seven days and consecrating the seventh day as the Sabbath. *Forty* is a special number, as in the forty days and forty nights that Moses spent upon Mount Sinai, or the forty years decreed for the Israelite wandering in the Sinai wilderness. These are numbers familiar to us from the Sumerian tales—the twelve of the solar system and the twelve-month calendar of Nippur; the seven as the planetary number of the Earth (when the Anunnaki counted from the outside in) and of Enlil as Earth's Commander; the forty as Ea/Enki's numerical rank.

The number *fifty* is also present. Fifty, as the reader knows, was a number with "sensitive" aspects—it was the original rank number of Enlil and the stand-in rank of his heir apparent, Ninurta; and more significantly, in the days of the Exodus, it connoted symbolism to Marduk and his Fifty Names. Extra attention is therefore called for when we find that "fifty" was granted extraordinary importance—*it was used to create a new Unit of Time,* the *fifty-year* **Jubilee.**

While the calendar of Nippur was clearly adopted as the calendar by which the festivals and other Israelite religious rites were to be observed, special regulations were dictated for the fiftieth year; it was given a special name, that of a *Jubilee* Year: "A hallowed Jubilee year it shall be unto you" (*Leviticus* Chapter 25). In such a year, unprecedented freedoms were to take place. The count was to be done by counting the New Year's Day of Atonements for seven years sevenfold, forty-nine times; then on the Day of Atonement on the year thereafter, the fiftieth year, the trumpet call of a *ram's horn* was to be sounded throughout the land, and freedom was to be proclaimed for the land and all who dwelled in it: people should return to their families; property should return to its original owners—all land and house sales shall be redeemable and undone; slaves (who had to be treated at all times as hired help!) shall be set free, and liberty shall be given the land itself by leaving it fallow that year.

As much as the concept of a "Year of Freedom" is novel and unique, the choice of fifty as a calendrical unit seems odd (we have adopted 100—a century—as a convenient unit of time). Then the name given to such a once-in-fifty year is even more intriguing. The word that is translated "Jubilee" is *Yovel* in the Hebrew

Bible, and it means "a ram." So one can say that what was decreed was a *"Year of the Ram,"* to repeat itself every fifty years, and to be announced by sounding the *ram's horn.* Both the choice of fifty for the new time unit and its name raise the unavoidable question: Was there a hidden aspect here, related to Marduk and his Age of the Ram?

Were the Israelites told to keep counting "fifty years" until some significant divine event, relating either to the Age of the Ram or to the holder of the Rank of Fifty—*when everything shall turn back to a new beginning?*

While no obvious answer is offered in those biblical chapters, one cannot avoid searching for clues by pursuing a significant and very similar year-unit on the other side of the world: not fifty, but fifty-two. It was the Secret Number of the Mesoamerican god Quetzalcoatl, who according to Aztec and Mayan legends gave them civilization, including their three calendars. In *The Lost Realms* we have identified Quetzalcoatl as the Egyptian god Thoth, whose secret number was fifty-two—a calendrical-based number, for it represented the fifty-two weeks of seven days in a solar year.

The oldest of the three Mesoamerican calendars is known as the Long Count: it counted the number of days from a "Day One" that scholars have identified as August 13, 3113 B.C.E. Alongside this continuous but linear calendar there were two cyclical calendars. One, the *Haab,* was a solar-year calendar of 365 days, divided into 18 months of 20 days each plus an additional 5 special days at year's end. The other was the *Tzolkin,* a Sacred Calendar of only 260 days, composed of a 20-day unit rotated 13 times. The two cyclical calendars were then meshed together, as two geared wheels **(Fig. 67)**, to create the Sacred Round of fifty-two years, when these two counts returned to their common starting point and started the counts all over again.

This "bundle" of fifty-two years was a most important unit

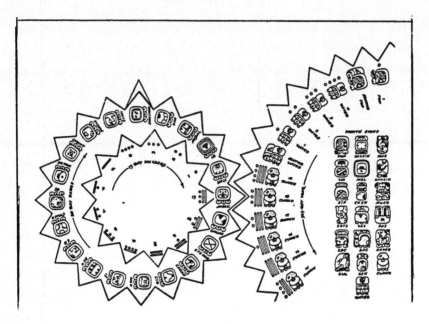

FIGURE 67

of time, because it was linked to the promise of Quetzalcoatl, who at some point left Mesoamerica, to return on his Sacred Year. The Mesoamerican peoples therefore used to gather on mountains every fifty-two years to expect the promised Return of Quetzalcoatl. (In one such Sacred Year, 1519 A.D., a white-faced and bearded Spaniard, Hernando Cortes, landed on Mexico's Yucatan coast and was welcomed by the Aztec king Montezuma as the returning god—a costly mistake, as we now know.)

In Mesoamerica, the "bundle year" served for a countdown to the promised "Year of Return," and the question is, *Was the "Jubilee year" intended to serve a similar purpose?*

Searching for an answer, we find that when the linear fifty-year time unit is meshed with the zodiacal cyclical unit of seventy-two—the time that a shift of one degree requires—we arrive at 3,600 (50 × 72 = 3,600), which was the (mathematical) orbital period of Nibiru.

By linking a Jubilee calendar and the zodiacal calendar

to Nibiru's orbit, was the biblical God saying, "When you enter the Promised Land, start the countdown to the Return"?

Some two thousand years ago, during a time of great messianic fervor, it was recognized that the Jubilee was a divinely inspired time unit for predicting the future—for calculating when the meshed geared wheels of time will announce the Return. That recognition underlies one of the most important postbiblical books, known as *The Book of Jubilees.*

Though available now only from its Greek and later translations, it was originally written in Hebrew, as fragments found among the Dead Sea scrolls confirm. Based on earlier extrabiblical treatises and sacred traditions, it rewrote the Book of Genesis and part of Exodus according to a calendar based on the Jubilee Time Unit. It was a product, all scholars agree, of messianic expectations at the time when Rome occupied Jerusalem, and its purpose was to provide a means by which to predict when the Messiah shall come—when the **End of Days** shall occur.

It is the very task we have undertaken.

Chapter X

THE CROSS ON THE HORIZON

About sixty years after the Israelites' Exodus, highly unusual religious developments took place in Egypt. Some scholars view those developments as an attempt to adopt Monotheism—perhaps under the influence of the revelations at Mount Sinai. What they have in mind is the reign of Amenhotep (sometimes rendered as Amenophis) IV who left Thebes and its temples, gave up the worship of Amon, and declared ATEN the sole creator god.

As we shall show, that was not an echo of Monotheism, but another harbinger of an expected Return—the return, into view, of the Planet of the Cross.

The Pharaoh in question is better known by the new name he had adopted—*Akhen-Aten* ("The servant/worshipper of Aten"), and the new capital and religious center that he had established, *Akhet-Aten* ("Aten of the Horizon"), is better known by the site's modern name, Tell el-Amarna (where the famed ancient archive of royal international correspondence was discovered).

Scion of Egypt's famed eighteenth Dynasty, Akhenaten reigned from 1379 to 1362 B.C.E., and his religious revolution did not last. The priesthood of Amon in Thebes led the opposition, presumably because it was deprived of its positions of power and wealth, but it is of course possible that the objections were genuinely on religious grounds, for Akhenaten's successors (of whom most famed was Tut-Ankh-Amen) resumed the inclusion of Ra/Amon in their theophoric names. No sooner was Akhenaten gone than the new capital, its temples, and its palace were torn down and systematically destroyed. Nevertheless, the remains that archaeologists have found throw enough light on Akhenaten and his religion.

The notion that the worship of the Aten was a form of monotheism—worship of a sole universal creator—stems primarily from some of the hymns to the Aten that have been found; they include such verses as "*O sole god*, like whom there is no other . . . The world came into being by thy hand." The fact that, in a clear departure from Egyptian customs, representation of this god in anthropomorphic form was strictly forbidden sounds very much like Yahweh's prohibition, in the Ten Commandments, against making any "graven images" to worship. Additionally, some portions of the Hymns to Aten read as if they were clones of the biblical Psalms—

> *O living Aten,*
> *How manifold are thy works!*
> *They are hidden from the sight of men.*
> *O sole god, beside whom there is no other!*
> *Thou didst create the earth according to thy desire*
> *whilst thou wast alone.*

The famed Egyptologist James H. Breasted (*The Dawn of Conscience*) compared the above verses to Psalm 104, beginning with verse 24—

> *O Lord, how manifold are thy works!*
> *In wisdom hast thou made them all;*
> *the Earth is full of thy riches.*

The similarity, however, arises not because the two, Egyptian hymn and biblical Psalm, copy each other, but because both speak of the same celestial god of the Sumerian Epic of Creation— of Nibiru—that shaped the Heavens and created the Earth, imparting to it the "seed of life."

Virtually every book on ancient Egypt will tell you that the "Aten" disc that Akhenaten made the central object of worship represented the benevolent Sun. If so, it was odd that in a marked departure from Egyptian temple architecture that oriented the temples to the solstices on a southeast-northwest axis, Akhenaten oriented his Aten temple on an east–west axis—but had it facing west, *away* from the Sun at sunrise. If he was expecting a celestial reappearance from a direction *opposite* to that of where the Sun rises, it could not be the Sun.

A close reading of the hymns reveals that Akhenaten's "star god" was not Ra as Amon "the Unseen," but a different kind of Ra: it was the celestial god who had "existed from primeval time . . . *The one who renews himself*" as it *reappears* in all its glory, a celestial god that was *"going afar and returning."* On a daily basis, those words could indeed apply to the Sun, but on a long-term basis, the description fitted Ra only as Nibiru: it did become unseen, the hymns said, because it was "far away in heaven," because it went "to the rear of the horizon, to the height of heaven." And now, Akhenaten announced, it was coming back in all its glory. Aten's hymns prophesied its reappearance, its return "beautiful on the horizon of heaven . . . Glittering, beautiful, strong," ushering *a time of peace and benevolence to all.* These words express clear messianic expectations that have nothing to do with the Sun.

In support of the "Aten is the Sun" explanation, various depictions of Akhenaten are offered; they show **(Fig. 68)** him and his wife blessed by, or praying to, a rayed star; it is the Sun, most Egyptologists say. The hymns do refer to the Aten as a manifestation of Ra, which to Egyptologists who have deemed Ra to be the Sun means that Aten, too, represented the Sun; but if Ra

FIGURE 68

was Marduk and the celestial Marduk was Nibiru, then Aten, too, represented Nibiru and not the Sun. Additional evidence comes from sky maps, some painted on coffin lids **(Fig. 69)**, that clearly showed the twelve zodiacal constellations, the rayed Sun, and other members of the solar system; but the planet of Ra, the "Planet of Millions of Years," is shown as an extra planet in its own *large separate celestial barque beyond the Sun*, with the pictorial hieroglyph for "god" in it—Akhenaten's "Aten."

What, then, was Akhenaten's innovation, or, rather, digression, from the official religious line? At its core his "transgression" was the same old debate that had taken place 720 years earlier about *timing*. Then the issue was: Has Marduk/Ra's time for supremacy come, has the Age of the Ram begun in the heavens? Akhenaten shifted the issue from Celestial Time (the zodiacal clock) to Divine Time (Nibiru's orbital time), changing the question to: *When will the **Unseen** celestial god reappear* and become visible—"beautiful on the horizon of heaven"?

His greatest heresy in the eyes of the priests of Ra/Amon can be judged by the fact that he erected a special monument honoring the *Ben-Ben*—an object that had been revered generations

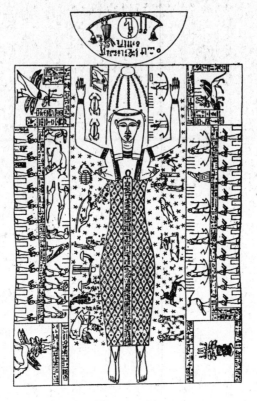

FIGURE 69

earlier as the vehicle in which Ra had arrived on Earth from the heavens **(Fig. 70)**. It was an indication, we believe, that what he was expecting in connection with Aten was a Reappearance, a Return not just of the Planet of the Gods, but another arrival, *a New Coming of the gods themselves!*

This, we must conclude, was the innovation, the difference introduced by Akhenaten. In defiance of the priestly establishment, and no doubt prematurely in their opinion, he was announcing the coming of a new messianic time. This heresy was aggravated by the fact that Akhenaten's pronouncements about the returning Aten were accompanied by a personal claim: Akhenaten increasingly referred to himself as *the god's prophet-son, one "who came forth from the god's body,"* and to whom alone the deity's plans were revealed:

FIGURE 70

There is no other that knoweth thee
except thy son Akhenaten;
Thou hast made him wise in thy plans.

And this, too, was unacceptable to the Theban priests of Amon. As soon as Akhenaten was gone (and it is uncertain how . . .), they restored the worship of Amon—the Unseen god—and smashed and destroyed all that Akhenaten had erected.

That the Aten episode in Egypt, as the introduction of the Jubilee—the "Year of the Ram"—were the stirrings of a wider expectation of a Return of a celestial "star god" is evident from yet another biblical reference to the Ram, yet another manifestation of a **Countdown to the Return.**

It is the record of an unusual incident at the end of the Exodus. It is a tale that is replete with puzzling aspects, and one that ends with a divinely inspired vision of things to come.

The Bible repeatedly declared divination by examining animal entrails, consulting with spirits, soothsaying, enchanting, conjuring, and fortune-telling to be "abominations unto Yahweh"—all manners of sorcery practiced by other nations that the Israelites must avoid. At the same time, it asserted—quoting Yahweh himself—that dreams, oracles, and visions could be legitimate ways of divine communication. It is such a distinction that explains why the *Book of Numbers* devotes three long chapters (22–24) to tell—approvingly!—the story of a non-Israelite Seer and Oracle-teller. His name was Bil'am, rendered Balaam in English Bibles.

The events described in those chapters took place when the Israelites ("Children of Israel" in the Bible), having left the Sinai Peninsula, went around the Dead Sea on the east, advancing northward. As they encountered the small kingdoms that occupied the lands east of the Dead Sea and the Jordan River, Moses sought from their kings permission for peaceful passage; it was, by and large, refused. The Israelites, having just defeated the Ammonites, who did not let them pass through peacefully, now "were encamped in the plains of Mo'ab, on the side of the Jordan that is opposite Jericho," awaiting the Moabite king's permission to pass through his land.

Unwilling to let "the horde" pass yet afraid to fight them, the king of Mo'ab—Balak the son of Zippor—had a bright idea. He sent emissaries to fetch an internationally renowned seer, Bala'am the son of Be'or, and have him "put a curse on these people for me," to make it possible to defeat and chase them away.

Balaam had to be entreated several times before he accepted the assignment. First at Balaam's home (somewhere near the Euphrates River?) and then on the way to Moab, an Angel of God (the word in Hebrew, *Mal'ach*, literally means "emissary") appears and gets involved in the proceedings; he is sometimes visible and sometimes invisible. The Angel allows Balaam to accept the assignment only after making sure that Balaam understands that he is to utter only divinely inspired omens. Puzzlingly, Balaam calls Yahweh "my God" when he repeats this condition, first to the king's ambassadors and then to the Moabite king himself.

A series of oracular settings are then arranged. The king takes Balaam to a hilltop from which he can see the whole Israelite encampment, and on the Seer's instructions he erects seven altars, sacrifices seven bullocks and seven rams, and awaits the oracle; but from Balaam's mouth come words not of accusation but of praise for the Israelites.

The persistent Moabite king then takes Balaam to another mount, from which just the edge of the Israelite encampment can be seen, and the procedure is repeated a seond time. But again Balaam's oracle blesses rather than curses the Israelites: I see them coming out of Egypt protected by a god with spreading ram's horns, he says—it is a nation destined for kingship, a nation that like a lion will arise.

Determined to try again, the king now takes Balaam to a hilltop that faces the desert, facing away from the Israelite encampment; "maybe the gods will let you proclaim curses there," he says. Seven altars are again erected, on which seven bullocks and seven rams are sacrificed. But Balaam now sees the Israelites and their future not with human eyes but "in a divine vision." For the second time he sees the nation protected, as it came out of Egypt, by a god with spreading rams' horns, and envisions Israel as a nation that "like a lion will arise."

When the Moabite king protests, Balaam explains that no matter what gold or silver he be offered, he can utter only the words that God puts in his mouth. So the frustrated king gives up and lets Balaam go. But now Balaam offers the king free advice: Let me tell you what the future holds, he says to the king—"that which will come about to this nation and to your people at the end of days"—and proceeds to describe the divine vision of the future by relating it to a "star":

> I see it, though not now;
> I behold it, though it is not near:
> A Star of Jacob is on its course.
> A Scepter from Israel will arise—
> Moab's quarters it will crush,
> all the Children of Seth it will unsettle.

NUMBERS 24: 17

Balaam then turned and cast his eyes toward the Edomites, Amalekites, Kenites, and other Canaanite nations, and pronounced an oracle thereon: Those who will survive the wrath of Jacob shall fall into the hands of Assyria; then Assyria's turn will come, and it shall forever perish. And having pronounced that oracle, "Balaam rose up and went back to his place; and Balak too went on his way."

Though the Balaam episode has naturally been the subject of discussion and debate by biblical and theological scholars, it remains baffling and unresolved. The text switches effortlessly between references to the *Elohim*—"gods" in the plural—and to Yahweh, the sole God, as the Divine Presence. It gravely transgresses the Bible's most basic prohibition by applying to the God who brought the Israelites out of Egypt a physical image, and then compounds the transgression by envisioning Him in the image of "a ram with spreading horns"—an image that has been the Egyptian depiction of Amon **(Fig. 71)**! The approving attitude toward a professional seer in a Bible that prohibited soothsaying, conjuring, and so on adds to the feeling that the whole tale was, originally, a non-Israelite tale, and yet the Bible incorporated it, devoting to it substantial space, so the incident and its message must have been considered a significant prelude to the Israelite possession of the Promised Land.

The text suggests that Balaam was an Aramaean, residing somewhere up the Euphrates River; his prophetic oracles expanded from the fate of the Children of Jacob to the place of Israel among the nations to oracles regarding the future of such other nations—even of distant and yet-to-come imperial Assyria. The oracles were thus an expression of wider non-Israelite expectations at the time. ***By including the tale, the Bible combined the Israelite destiny with Mankind's universal expectations.***

FIGURE 71

Those expectations, the Balaam tale indicates, were chan-
neled along two paths—the zodiacal cycle on the one hand, and
the Returning Star's course on the other hand.

The zodiacal references are strongest regarding the Age of the
Ram (and its god!) at the time of the Exodus, and become oracular and prophetic as the Seer Balaam envisions the Future, when
the zodiacal constellation symbols of the Bull and the Ram ("bullocks and rams for sevenfold sacrifices") and the Lion ("when the
Royal Trumpet shall be heard in Israel") are invoked (*Numbers*
Chapter 23). And it is when envisioning that Distant Future that
the Balaam text employs the significant term *At the End of Days*

as the time to which the prophetic oracles apply (*Numbers* 24: 14).

The term directly links these non-Israelite prophecies to the destiny of Jacob's offspring because it was used by Jacob himself as he lay on his deathbed and gathered his children to hear oracles regarding their future (*Genesis* Chapter 49). "Come gather together," he said, "that I may tell you that which shall befall you at *the End of Days*." The oracles, individually stated for each one of the twelve future Tribes of Israel, are deemed by many to be related to the twelve zodiacal constellations.

And what about the Star of Jacob—an explicit vision by Balaam?

In scholarly biblical discussions, it is usually considered in an astrological rather than an astronomical context at best, and more often than not, the tendency has been to deem the reference to "Jacob's Star" as purely figurative. But what if the reference was indeed to a "star" orbiting on its course—a planet prophetically seen though it is not yet visible?

What if Balaam, like Akhenaten, was speaking of the return, the reappearance, of Nibiru? Such a return, it must be realized, would be an extraordinary event that occurs once in several millennia, an event that had repeatedly signified the most profound watersheds in the affairs of gods and men.

This is not just a rhetorical question. In fact, the unfolding events were increasingly indicating that an overwhelmingly significant occurrence was in the offing. Within a century or so of the preoccupations and predictions regarding the Returning Planet that we find in the tales of the Exodus, Balaam, and Akhenaten's Egypt, Babylon itself started to provide evidence of such wide-spreading expectations, and the most prominent clue was the **Sign of the Cross.**

In Babylon, the time was that of the Kassite dynasty, of which we have written earlier. Little has remained of their reign in Babylon itself, and as stated earlier those kings did not excel in keeping royal records. But they did leave behind telltale depictions—and international correspondence of letters on clay tablets.

It was in the ruins of Akhet-Aten, Akhenaten's capital—a site now known as Tell el-Amarna in Egypt—that the famed "el-Amarna Tablets" were discovered. Of the 380 clay tablets, all except three were inscribed in the Akkadian language, which was then the language of international diplomacy. While some of the tablets represented copies of royal letters sent from the Egyptian court, the bulk were original letters received from foreign kings.

The cache was the royal diplomatic archive of Akhenaten, and the tablets were predominantly correspondence he had received from the kings of Babylon!

Did Akhenaten use those exchanges of letters with his counterparts in Babylon to tell them of his newfound Aten religion? We really don't know, for all we have are a Babylonian king's letters to Akhenaten in which he complained that gold sent to him was found short in weight, that his ambassadors were robbed on the way to Egypt, or that the Egyptian king failed to inquire about his health. Yet the frequent exchanges of ambassadors and other emissaries, even offers of intermarriage, as well as the calling of the Egyptian king "my brother" by the Babylonian king, must lead to a conclusion that the hierarchy in Babylon was fully aware of the religious goings-on in Egypt; and if Babylon wondered, "What is this 'Ra as a Returning Star' commotion?" Babylon must have realized that it was a reference to "Marduk as a Returning Planet"—*to Nibiru orbiting back.*

With the tradition of celestial observations so much older and more advanced in Mesopotamia than in Egypt, it is of course possible that the royal astronomers of Babylon had come to conclusions regarding Nibiru's return without Egyptian aid, and even ahead of the Egyptians. Be that as it may, it was in the thirteenth century B.C.E. that the Kassite kings of Babylon started to signal, in a variety of ways, their own fundamental religious changes.

In **1260 B.C.E.** a new king ascended the throne in Babylon and adopted the name Kadashman-Enlil—a theophoric name surprisingly venerating Enlil. It was no passing gesture, for he was followed on the throne, for the next century, by Kassite kings bearing theophoric names venerating not only Enlil but also Adad—a surprising gesture suggesting a desire for divine reconciliation. That something unusual was expected was further evidenced on commemorative monuments called *kudurru*—"rounded stones"—that were set up as boundary markers. Inscribed with a text stating the terms of the border treaty (or land grant) and the

FIGURE 72

oaths taken to uphold it, the kudurru was sanctified by symbols of the celestial gods. The divine zodiacal symbols—all twelve of them—were frequently depicted **(Fig. 72)**; orbiting above them

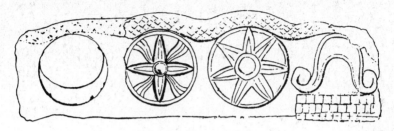

FIGURE 73

were the emblems of the Sun, the Moon, and Nibiru. In another depiction **(Fig. 73)**, Nibiru was shown in the company of Earth (the seventh planet) and the Moon (and the umbilical-cutter symbol for Ninmah).

Significantly, Nibiru was depicted no longer by the Winged Disc symbol, but rather in a new way—as the planet of the **radiating cross**—befitting its description by the Sumerians in the "Olden Days" as a radiating planet about to become the "Planet of the Crossing."

This way of showing a long-unobserved Nibiru by a symbol of a radiating cross began to become more common, and soon the Kassite kings of Babylon simplified the symbol to just a **Sign of the Cross,** replacing with it the Winged Disc symbol on their royal seals **(Fig. 74)**. This cross symbol, which looks like the much later Christian "Maltese Cross," is known in studies of ancient glyptic as a "Kassite Cross." As another depiction indicates, the symbol of the cross was for a planet *clearly not the same as the Sun,* which is *separately shown* along with the Moon-crescent and the six-pointed star symbol for Mars **(Fig. 75)**.

As the first millenium B.C.E. began, Nibiru's Sign of the Cross spread from Babylonia to seal designs in nearby lands. In the

FIGURE 74

absence of Kassite religious or literary texts, it is a matter of conjecture what messianic expectations might have accompanied these changes in depictions. Whatever they were, they intensified the ferocity of the attacks by the Enlilite states—Assyria, Elam—on Babylon and their opposition to Marduk's hegemony. Those attacks delayed, but did not prevent, the eventual adoption of the Sign of the Cross in Assyria itself. As royal monuments reveal, it was worn, most conspicuously, by Assyria's kings on their

FIGURE 75

chests, near their hearts **(Fig. 76)**—the way devout Catholics wear the cross nowadays. Religiously and astronomically, it was a most significant gesture. That it was also a widespread manifesta-

FIGURE 76

tion is suggested by the fact that in Egypt, too, depictions were found of a king-god wearing, like his Assyrian counterparts, the sign of the cross on his chest **(Fig. 77)**.

The adoption of the Sign of the Cross as the emblem of Nibiru, in Babylon, Assyria, and elsewhere, was not a surprising innovation. The sign had been used before—by the Sumerians and Akkadians. *"Nibiru—let 'Crossing' be its name!"* the Epic of Creation stated; and accordingly its symbol, the cross, had been employed in Sumerian glyptic to denote Nibiru, but then *it always signified its **Return into visibility**.*

FIGURE 77

Enuma elish, the Epic of Creation, clearly stated that after the Celestial Battle with Tiamat, the Invader made a grand orbit around the Sun and returned to the scene of the battle. Since Tiamat orbited the Sun in a plane called the Ecliptic (as other members of our Sun's planetary family do), it is to that place in the heavens that the Invader had to return; and when it does so, orbit after orbit after orbit, it is there that it *crosses the plane of the ecliptic.* A simple way to illustrate this would be to show the orbital path of the well-known Halley's Comet **(Fig. 78),** which emulates on a greatly reduced scale the orbit of Nibiru: its inclined orbit brings it, as it nears the Sun, from the south, from below the ecliptic, near Uranus. It arches above the ecliptic and makes the turn around the Sun, saying "Hello" to Saturn, Jupiter, and Mars; then it comes down and crosses the ecliptic near the site of Nibiru's Celestial Battle with Tiamat—the Crossing (marked "**X**")—and is gone, only to come back as its orbital Destiny prescribes.

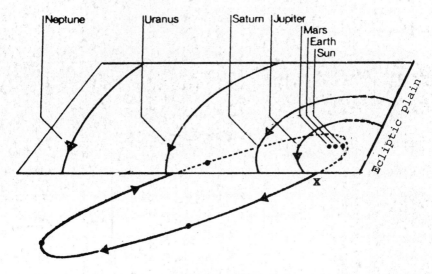

FIGURE 78

That point, in the heavens and in time, is **The Crossing**—it is then, *Enum elish* stated, that the planet of the Anunnaki becomes the **Planet of the Cross:**

> *Planet NIBIRU:*
> *The Crossroads of Heaven and Earth it shall occupy . . .*
> *Planet NIBIRU:*
> *The central position he holds . . .*
> *Planet NIBIRU:*
> *It is he who without tiring*
> *the midst of Tiamat keeps crossing;*
> **Let "Crossing" be his name!**

Sumerian texts dealing with landmark events in Mankind's saga provide specific indications regarding the periodic appearances of the Planet of the Anunnaki—approximately every 3,600 years— and always at crucial junctions in Earth's and Mankind's history. It was at such times that the planet was called Nibiru, and its glyptical depictions—even in early Sumerian times—were the Cross.

That record began with the Deluge. Several texts dealing with the Deluge associated the watershed catastrophe with the appearance of the celestial god, Nibiru, in the Age of the Lion (circa 10,900 B.C.E.)—it was "the constellation of the Lion that measured the waters of the deep," one text said. Other texts described the appearance of Nibiru at Deluge time as a radiating star, and depicted it accordingly **(Fig. 79)**—

FIGURE 79

When they shall call out "Flooding!"
It is the god Nibiru . . .
*Lord whose **shining crown** with terror is laden;*
*Daily within the Lion **he is afire**.*

The planet returned, reappeared, and again became "Nibiru" when Mankind was granted farming and husbandry, in the mid-eighth millennium B.C.E.; depictions (on cylinder seals) illustrating the beginning of agriculture used the Sign of the Cross to show Nibiru visible in Earth's skies **(Fig. 80)**.

Finally and most memorably for the Sumerians, the planet was visible once again when Anu and Antu came to Earth on a state visit circa 4000 B.C.E., in the Age of the Bull (Taurus). The city that was later known for millennia as Uruk was established in their honor, a ziggurat was erected, and from its stages the appearance of the planets on the horizon, as the night sky darkened, was observed. When Nibiru came into view, a shout went up:

FIGURE 80

"The Creator's image has arisen!" and all present broke into hymnal songs of praise for "the planet of the Lord Anu."

Nibiru's appearance at the start of the Age of the Bull meant that at the time of heliacal rising—when dawn begins but the horizon is still dark enough to see the stars—the constellation in the background was that of Taurus. But the fast-moving Nibiru, arcing in the skies as it circled the Sun, soon descended back to cut across the planetary plain ("ecliptic") to the point of Crossing. There the crossing was observed against the background of the constellation of the Lion. Several depictions, on cylinder seals and in astronomical tablets, used the cross symbol to indicate Nibiru's arrival when Earth was in the Age of the Bull and its crossing was observed in the constellation of the Lion (cylinder seal depiction, **Fig. 81**, and as illustrated in **Fig. 82**).

FIGURE 81

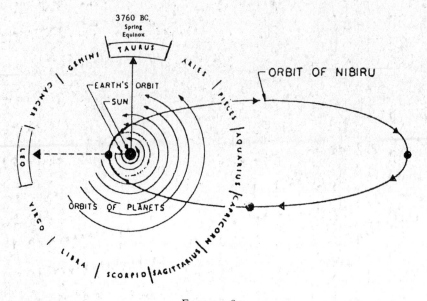

FIGURE 82

The change from the Winged Disc symbol to the Sign of the Cross thus was not an innovation; it was reverting to the way in which the Celestial Lord had been depicted in earlier times—but only when in its great orbit it crossed the ecliptic and became "Nibiru."

As in the past, the renewed display of the Sign of the Cross signified reappearance, coming back into view, RETURN.

Chapter XI

The Day of the Lord

As the last millennium B.C.E. began, the appearance of the Sign of the Cross was a harbinger of the Return. It was also then that a temple to Yahweh in Jerusalem forever linked its sacred site to the course of historic events and to Mankind's messianic expectations. The time and the place were no coincidence: the impending Return dictated the enshrinement of the erstwhile Mission Control Center.

Compared to the mighty and conquering imperial powers of those days—Babylonia, Assyria, Egypt—the Hebrew kingdom was a midget. Compared to the greatness of their capitals—Babylon, Nineveh, Thebes—with their sacred precincts, ziggurats, temples, processional ways, ornate gates, majestic palaces, hanging gardens, sacred pools, and river harbors—Jerualem was a small city with hastily built walls and an iffy water supply. And yet, millennia later, it is Jerusalem, a living city, that is in our hearts and in the daily headlines, while the grandeur of the other nations' capitals has turned to dust and crumbled ruins.

What made the difference? The Temple of Yahweh that was built in Jerusalem, and its Prophets whose oracles came

true. **Their prophecies, one therefore believes, still hold the key to the Future.**

The Hebrew association with Jerusalem, and in particular with Mount Moriah, goes back to the time of Abraham. It was when he had fulfilled his assignment of protecting the spaceport during the War of the Kings that he was greeted by Malkizedek, the king of *Ir-Shalem* (Jerusalem), "who was a priest of the God Most High." There Abraham was blessed, and in turn took an oath, "by the God Most High, possessor of Heaven and Earth." It was again there, when Abraham's devotion was tested, that he was granted a Covenant with God. Yet it took a millennium, until the right time and circumstances, for the Temple to be built.

The Bible asserted that the Jerusalem temple was unique—and so indeed it was: it was conceived to preserve the "Bond Heaven-Earth" that the DUR.AN.KI of Sumer's Nippur had once been.

> *And it came to pass*
> *in the fourhundred and eightieth year*
> *after the Children of Israel came out of Egypt,*
> *in the fourth year of Solomon's reign,*
> *in the second month,*
> *that he began to build the House of the Lord.*

Thus does the Bible record, in the first *Book of Kings* (6: 1), the memorable start of the construction of the Temple of Yahweh in Jerusalem by King Solomon, giving us the exact date of the event. It was a crucial, decisive step whose consequences are still with us; and **the time, it must be noted, was when Babylon and Assyria adopted the Sign of the Cross as the harbinger of the Return . . .**

The dramatic story of the Jerusalem Temple starts not with Solomon but with King David, Solomon's father; and how he happened to become Israel's king is a tale that reveals a divine plan: **to prepare for the Future by resurrecting the Past.**

David's legacy (after a reign of 40 years) included a greatly expanded realm, reaching in the north as far as Damascus (and including the Landing Place!), many magnificent Psalms, and the groundwork for Yahweh's temple. Three divine emissaries played key roles in the making of this king and his place in history; the Bible lists them as "Samuel the Seer, Nathan the Prophet, and Gad the Visionary." It was Samuel, the custodian-priest of the Ark of the Covenant, who was instructed by God to "take the youth David, son of Jesse, from herding sheep to be shepherd of Israel," and Samuel "took the oil-filled horn and *anointed him* to reign over Israel."

The choosing of the young David, who was shepherding his father's flock, to be shepherd over Israel was doubly symbolic, for it harks back to the golden age of Sumer. Its kings were called LU.GAL, "Great Man," but they strove to earn the cherished title EN.SI, "Righteous Shepherd." That, as we shall see, was only the beginning of David's and the Temple's links to the Sumerian past.

David began his reign in Hebron, south of Jerusalem, and that, too, was a choice filled with historic symbolism. The previous name of Hebron, the bible repeatedly pointed out, was *Kiryat Arba*, "the fortified city of Arba." And who was Arba? "He was a *Great Man* of the *Anakim*"—two biblical terms that render in Hebrew the Sumerian LU.GAL and ANUNNAKI. Starting with passages in the book of *Numbers,* and then in *Joshua, Judges,* and *Chronicles,* the Bible reported that Hebron was a center of the descendants of the "Anakim, who as the Nefilim are counted," thus connecting them to the Nefilim of *Genesis 6* who intermarried with the Daughters of Adam. Hebron was still inhabited at the time of the Exodus by three sons of Arba, and it was Caleb the son of Jephoneh who captured the city and slew them in behalf of Joshua. *By choosing to be king in Hebron, David established his kingship as a direct continuation of kings linked to the Anunnaki of Sumerian lore.*

He reigned in Hebron for seven years, and then moved his capital to Jerusalem. His seat of kingship—the "City of David"—was built on Mount Zion, just south of and separated by a small valley from Mount Moriah (where the platform built by the Anunnaki was, **Fig. 83**). He constructed the *Miloh*, the Filling,

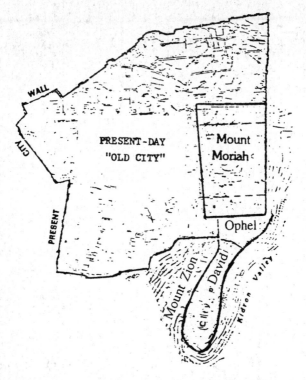

FIGURE 83

to close the gap between the two mounts, as a first step to building, on the platform, Yahweh's temple; but all he was allowed to erect on Mount Moriah was an altar. God's word, through the Prophet Nathan, was that because David had shed blood in his many wars, not he but his son Solomon would build the temple.

Devastated by the prophet's message, David went and "sat before Yahweh," in front of the Ark of the Covenant (which was still housed in a portable tent). Accepting God's decision, he asked for one reward for his devout loyalty to Him: an assurance, a sign, that it would indeed be the House of David that would build the Temple and be forever blessed. That very night, sitting in front of the Ark of the Covenant by which Moses had communicated with the Lord, he received a divine sign: he was given *a Tavnit*—*a scale model*—of the future temple!

One can shrug off the tale's veracity were it not for the fact that what happened that night to King David and his temple project was the equivalent of the *Twilight Zone* tale of the Sumerian king Gudea, who more than a thousand years earlier was likewise given in a vision-dream a tablet with the architectural plan and a brick mold for the construction of a temple in Lagash for the god Ninurta.

When he neared the end of his days, King David summoned to Jerusalem all the leaders of Israel, including the tribal chiefs and the military commanders, the priests and the royal office holders, and told them of Yahweh's promise; and in full view of those gathered he handed to his son Solomon "the *Tavnit* of the temple and all its parts and chambers . . . the Tavnit that he received by the Spirit." There was more, for David also handed over to Solomon "all that Yahweh, in His own hand written, gave to me for understanding the workings of the Tavnit": A set of accompanying instructions, divinely written (*I Chronicles,* Chapter 28).

The Hebrew term *Tavnit* is translated in the King James English Bible "pattern" but is rendered "plan" in more recent translations, suggesting that David was given some kind of an architectural drawing. But the Hebrew word for "plan" is *Tokhnit. Tavnit*, on the other hand, is derived from the root verb that means "to construct, to build, to erect," so what David was given and what he handed over to his son Solomon was a "constructed model"—in today's parlance, a scale model. (Archaeological finds throughout the ancient Near East have indeed unearthed scale models of chariots, wagons, ships, workshops, and even multilevel shrines.)

The biblical books of *Kings* and *Chronicles* provide precise measurements and clear structural details of the Temple and its architectural designs. Its axis ran east–west, making it an "eternal temple" aligned to the equinox. Consisting of three parts (see Fig. 64), it adopted the Sumerian temple plans of a forepart (*Ulam* in Hebrew), a great central hall (*Hekhal* in Hebrew, stemming from the Sumerian E.GAL, "Large Abode"), and a Holy of Holies for the Ark of the Covenant. That innermost section was called the *Dvir* (the 'Speaker')—for it was by means of the Ark of the Covenant that God spoke to Moses.

As in Sumerian ziggurats, which traditionally were built to express the sexagesimal's "base sixty" concept, the Temple of Solomon also adopted sixty in its construction: the main section (the Hall) was 60 cubits (about 100 feet) in length, 20 cubits (60:3) wide, and 120 (60 × 2) cubits in height. The Holy of Holies was 20 by 20 cubits—just enough to hold the Ark of the Covenant with the two golden Cherubim atop it ("their wings touching"). Tradition, textual evidence, and archaeological research indicate that the Ark was placed precisely on the extraordinary rock on which Abraham was ready to sacrifice his son Isaac; its Hebrew designation, *Even Shatiyah*, means "Foundation Stone," and Jewish legends hold that it is from it that the world will be re-created. Nowadays it is covered over and surrounded by the Dome of the Rock **(Fig. 84)**. (Readers can find more about the sacred rock and its enigmatic cave and secret subterranean passages in *The Earth Chronicles Expeditions*.)

Though these were not monumental measurements compared to the skyscraping ziggurats, the Temple, when completed, was truly magnificent; it was also unlike any other contemporary temple in that part of the world. No iron or iron tools were used for its erection upon the platform (and absolutely none in its operation—all the utensils were of copper or bronze), and *the*

FIGURE 84

building was inlaid inside with gold; even the nails holding the golden plates in place were made of gold. The quantities of gold used (just "for the Holy of Holies, 600 talents; for the nails, fifty shekels") were enormous—so much so that Solomon arranged for special ships to bring gold from Ophir (believed to be in southeast Africa).

The Bible offers no explanation, neither for the prohibition against using anything made of iron on the site nor for the inlaying of everything inside the temple with gold. One can only speculate that iron was shunned because of its magnetic properties, and gold because it is the best electrical conductor.

It is significant that the only two other known instances of shrines so inlaid with gold are on the other side of the world. One is the great temple in Cuzco, the Inca capital in Peru, where the great god of South America, Viracocha, was worshipped. It was called the *Coricancha* ("Golden Enclosure"), for its Holy of

Holies was completely inlaid with gold. The other is in Puma-Punku on the shores of Lake Titicaca in Bolivia, near the famed ruins of Tiwanaku. The ruins there consist of the remains of four chamberlike stone buildings whose walls, floors, and ceilings were each cut out of a single colossal stone block. The four enclosures were completely inlaid inside with golden plates that were held in place with golden nails. Describing the sites (and how they were looted by the Spaniards) in *The Lost Realms*, I have suggested that Puma-Punku was erected for the stay of Anu and Antu when they visited Earth circa 4000 B.C.E.

According to the Bible, tens of thousands of workmen were needed for seven years for the immense undertaking. What, then, was the purpose of this House of the Lord? When all was ready, with much pomp and circumstance, the Ark of the Covenant was carried by priests and placed in the Holy of Holies. As soon as the Ark was put down and the curtains separating the Holy of Holies from the great hall were drawn, "the House of the Lord was filled with a cloud and the priests could not remain standing." Then Solomon offered a thanksgiving prayer, saying:

> *Lord who has chosen to dwell in the cloud:*
> *I have built for Thee a stately House,*
> *a place where you may dwell forever . . .*
> *Though the uttermost heavens cannot contain Thee,*
> *May you hear our supplications from Thine seat in heaven.*

"And Yahweh appeared to Solomon that night, and said to him: I have heard your prayer; I have chosen this site for my house of worship . . . From heaven I will hear the prayers of my people and forgive their transgressions . . . Now I have chosen and consecrated this House for my *Shem* to remain there forever" (*II Chronicles*, Chapters 6–7).

The word *Shem*—here and earlier, as in the opening verses of chapter 6 of *Genesis*—is commonly translated "Name." As far back as in my first book, *The Twelfth Planet*, I have suggested that the term originally and in the relevant context referred to what the Egyptians called the "Celestial Boat" and the Sumerians called MU—"sky ship"—of the gods. Accordingly, the Temple in Jerusalem, built atop the stone platform, with the Ark of the Covenant placed upon the sacred rock, was to serve as an earthly bond with the celestial deity—both for communcating and for the landing of his sky ship!

Throughout the Temple there was no statue, no idol, no graven image. The only object within it was the hallowed Ark of the Covenant—and "there was nothing in the Ark except the two tablets that were given to Moses in Sinai."

Unlike the Mesopotamian ziggurat temples, from Enlil's in Nippur to Marduk's in Babylon, this one was not a place of residence for the deity, where the god lived, ate, slept, and bathed. **It was a House of Worship, a place of divine contact; it was a temple for a Divine Presence by the Dweller in the Clouds.**

It is said that a picture is worth a thousand words; it is certainly true where there are few pertinent words but many relevant pictures.

It was about the time that the Jerusalem temple was completed and consecrated to the Dweller in the Clouds that a noticeable change in the sacred glyptic—the depiction of the divine—took place where such depictions were common and permissible, and (at the time) first and foremost in Assyria. They showed, most clearly, the god Ashur as a "dweller of the clouds," full face or with just his hand showing, frequently depicted holding a bow **(Fig. 85)**—a depiction reminding one of the Bible's tale of the Bow in the Cloud that was a divine sign in the aftermath of the Deluge.

FIGURE 85

FIGURE 86a

A century or so later, Assyrian depictions introduced a new variant of the God in the Cloud. Classified as "Deity in a Winged Disc," they clearly showed a deity inside the emblem of the Winged Disc, by itself **(Fig. 86a)** or as it joins the Earth (seven dots) and the Moon (crescent) **(Fig. 86b)**. Since the Winged Disc represented Nibiru, it had to be a deity *arriving with Nibiru*. Clearly, then, ***these depictions implied expectations of the nearing arrival not only of the planet, but also of its divine dwellers, pobably led by Anu himself.***

The changes in glyphs and symbols, begun with the Sign of the Cross, were manifestations of more profound expectations, of

FIGURE 86b

overwhelming changes and wider preparations called for by the expected Return. However, the expectations and preparations were not the same in Babylon as in Assyria. In one, the messianic expectations were centered on the god(s) who were already there; in the other, the expectations related to the god(s) about to return and reappear.

In Babylon the expectations were mostly religious—a messianic revival by Marduk through his son Nabu. Great efforts were undertaken to resume, circa **960 B.C.E.**, the sacred *Akitu* ceremonies in which the revised *Enuma elish*—appropriating to Marduk the creation of Earth, the reshaping of the Heavens (the Solar System), and the fashioning of Man—was publicly read. The arrival of Nabu from his shrine in Borsippa (just south of Babylon) to play a crucial role in the ceremonies was an essential part of the revival. Accordingly, the Babylonian kings who reigned between 900 B.C.E. and 730 B.C.E. resumed bearing Marduk-related names and, in great numbers, Nabu-related names.

The changes in Assyria were more geopolitical; historians consider the time—circa **960 B.C.E.**—as the start of the Neo-Assyrian Imperial period. In addition to inscriptions on monuments and palace walls, the main source of information about

Assyria in those days is the annals of its kings, in which they recorded what they did, year by year. Judging by that, their main occupation was Conquest. With unparalleled ferocity, its kings set out on one military campaign after another not only to have dominion over the olden Sumer & Akkad, but also over what they deemed essential for the Return: *Control of the space-related sites.*

That this was the purpose of the campaigns is evident not only from their targets, but also from the grand stone reliefs on the walls of Assyrian palaces from the ninth and eighth centuries B.C.E. (which one can see in some of the world's leading museums): as on some cylinder seals, they show the king and the high priest, accompanied by winged *Cherubim*—Anunnaki "astronauts"—flanking the Tree of Life as they welcome the coming of the god in the Winged Disc **(Fig. 87a,b)**. *A divine arrival was clearly expected!*

Historians connect the start of this Neo-Assyrian period to the establishment of a new royal dynasty in Assyria, when Tiglath-Pileser II ascended the throne in Nineveh. The pattern of aggrandizement at home and conquest, destruction, and annexation

FIGURE 87a

FIGURE 87b

abroad was set by that king's son and grandson, who followed him as kings of Assyria. Interestingly, their first target was the area of the Khabur River, with its important trade and religious center—Harran.

Their successors took it from there. Frequently bearing the same name as previous glorified kings (hence the numerations I, II, III, etc. for them), the successive kings expanded Assyrian control in all directions, but with special emphasis on the coastal cities and mountains of *La-ba-an* (Lebanon). Circa **860 B.C.E.** Ashurnasirpal II—who wore the cross symbol on his chest (see Fig. 76)—boasted of capturing the Phoenician coastal cities of Tyre, Sidon, and Gebal (Byblos), and of ascending the Cedar Mountain with its sacred site, the olden Landing Place of the Anunnaki.

His son and successor Shalmaneser III recorded the erecting there of a commemorative stela calling the place *Bit Adini*. The name literally meant "the Eden Abode"—and was known by that same name to the biblical Prophets. The Prophet Ezekiel castigated the king of Tyre for deeming himself a god because he had been to that sacred place and "moved within its fiery stones"; and the Prophet Amos listed it when he spoke of the coming **Day of the Lord**.

As could be expected, the Assyrians then turned their attention to the other space-related site. After the death of Solomon his kingdom was split by his contending heirs into "Judea" (with Jerusalem as capital) in the south and "Israel" and its ten tribes in the north. In his best-known inscribed monument, the Black Obelisk, Shalmaneser III recorded the receipt of tribute from the Israelite king Jehu and, in a scene dominated by the Winged Disc emblem of Nibiru, depicted him kneeling in obeisance **(Fig. 88)**. Both the Bible and the Assyrian annals recorded the

FIGURE 88

subsequent invasion of Israel by Tiglath-Pileser III (744–727 B.C.E.), the detaching of its better provinces, and the partial exile of its leaders. Then, in 722 B.C.E., his son Shalmaneser V overran what was left of Israel, exiled all of its people, and replaced them with foreigners; the Ten Tribes were gone, their wherabouts remaining a lasting mystery. (Why and how, on his return from Israel,

Shalmaneser was punished and abruptly replaced on the throne by another son of Tiglath-Pileser is also an unsolved mystery.)

Having already captured the Landing Place, the Assyrians were now at the doortsep of the final prize, Jerusalem; but again they held off the final assault. The Bible explained it by attributing it all to the will of Yahweh; an examination of Assyrian records suggests that what and when they did in Israel and Judea was synchronized with what and when they did about Babylon and Marduk.

After the capture of the space-related site in Lebanon—but before launching the campaigns toward Jerusalem—the Assyrians took an unprecedented step for reconciliation with Marduk. In 729 B.C.E. Tiglath-Pileser III entered Babylon, went to its sacred precinct, and "took the hands of Marduk." It was a gesture with great religious and diplomatic significance; the priests of Marduk approved the reconciliation by inviting Tiglath-Pileser to share in the god's sacramental meal. Following that, Tiglath-Pileser's son Sargon II marched southward into the olden Sumer & Akkad areas, and after seizing Nippur turned back to enter Babylon. In 710 B.C.E. he, like his father, "took the hands of Marduk" during the New Year ceremonies.

The task of capturing the remaining space-related site fell to Sargon's successor, Sennacherib. The assault on Jerusalem in 704 B.C.E., at the time of its King Hezekiah, is amply recorded both in Sennacherib's annals and in the Bible. But while Sennacherib in his inscriptions spoke just of the successful seizing of Judean provincial cities, the Bible provides a detailed tale of the siege of Jerusalem by a mighty Assyrian army that was miraculously wiped out by Yahweh's will.

Encircling Jerusalem and entrapping its people, the Assyrians engaged in psychological warfare by shouting discouraging words to the defenders on the city's walls, ending with vilification of Yahweh. The shocked king, Hezekiah, tore his clothes in mourning and prayed in the Temple to "Yahweh, the God of Israel, who rests upon the Cherubim, the sole God upon all the nations," for help. In response, the Prophet Isaiah conveyed to him God's oracle: the Assyrian king shall never enter the city, he will return home in failure, and there he will be assassinated.

> *And it came to pass that night*
> *that the Angel of Yahweh went forth*
> *and smote in the camp of the Asssyrians*
> *a hundred and eighty five thousand.*
> *And at sunrise, lo and behold,*
> *they were all dead corpses.*
> *So Sennacherib, the king of Assyria, departed*
> *and journeyed back to his abode in Nineveh*
>
> 2 KINGS 19: 35–36

To make sure the reader realizes that the whole prophecy came true, the biblical narrative then continues: "And Sennacherib went away, and journeyed back to Nineveh; and it was when he was bowing down in his temple to his god . . . that Adramelekh and Sharezzer struck him down with a sword, and they fled to the land of Ararat. His son Esarhaddon became king in his stead."

The biblical postscript is an amazingly informed record: Sennacherib was indeed murdered, by his own sons, in 681 B.C.E. For the second time, Assyrian kings who attacked Israel or Judea were dead as soon as they went back.

While prophecy—the foretelling of what is yet to happen—is inherently what is expected of a prophet, the Prophets of the Hebrew Bible were more than that. From the very beginning, as was made clear in *Leviticus,* a prophet was not to be "a magician, a wizard, an enchanter, a charmer or seer of spirits, a fortune-teller, or one who conjures the dead"—a pretty comprehensive list of the varied fortune-tellers of the surrounding nations. Their mission as *Nabih*—"Spokesmen"—was to convey to kings and peoples

Yahweh's own words. And as Hezekiah's prayer made clear, while the Children of Israel were His Chosen People, He was "sole God *upon all the nations.*"

The Bible speaks of prophets from Moses on, but only fifteen of them have their own books in the Bible. They include the three "majors"—Isaiah, Jeremiah, and Ezekiel—and twelve "minors." Their prophetic period began with Amos in Judea (circa **760 B.C.E.**) and Hoseah in Israel (750 B.C.E.) and ended with Malachi (circa 450 B.C.E.). As expectations of the Return took shape, geopolitics, religion, and actual happenings combined to serve as the foundation of biblical Prophecy.

The biblical Prophets served as Keepers of the Faith and were the moral and ethical compass of their own kings and people; they were also observers and predictors on the world arena by possessing uncannily accurate knowledge of goings-on in distant lands, of court intrigues in foreign capitals, of which gods were worshipped where, plus amazing knowledge of history, geography, trade routes, and military campaigns. *They then combined such awareness of the Present with knowledge of the Past to foretell the Future.*

To the Hebrew Prophets, Yahweh was not only *El Elyon*—"God Supreme"—and not only God of the gods, *El Elohim*, but a Universal God—of all nations, of the whole Earth, of the universe. Though His abode was in the Heaven of Heavens, He cared for his creation—Earth and its people. Everything that has happened was by His will, and His will was carried out by Emissaries—be it Angels, be it a king, be it a nation. Adopting the Sumerian distinction between predetermined Destiny and free-willed Fate, the Prophets believed that the Future could be foretold because it was all preplanned, yet on the way thereto, things could change. Assyria, for example, was at times called God's "rod of wrath" with which other nations were punished, but when it chose to act unnecessarily brutally or out of bounds, Assyria itself was in turn subjected to punishment.

The Prophets seemed to be delivering a two-track message not only in regard to current events, but also in respect to the Future.

Isaiah, for example, prophesied that Mankind should expect a Day of Wrath when all the nations (Israel included) shall be judged and punished—as well as look forward to an idyllic time when the wolf shall dwell with the lamb, men shall beat their swords into plowshares, and Zion shall be a light unto all nations.

The contradiction has baffled generations of biblical scholars and theologians, but a close examination of the Prophets' words leads us to an astounding finding: the Day of Judgment was spoken of as the **Day of the Lord**; the messianic time was expected at the **End of Days**; and the two were neither synonymous nor predicted as concurrent events. They were two separate events, due to occur at different times:

One, the Day of the Lord, a day of God's judgment, was about to happen;

The other, ushering a benevolent era, was yet to come, sometime in the future.

Did the words spoken in Jerusalem echo the debates in Nineveh and Babylon regarding which time cycle applies to the future of gods and men—Nibiru's orbital Divine Time or the zodiacal Celestial Time? Undoubtedly, as the eighth century B.C.E. was ending, it was clear in all three capitals that the two time cycles were not identical; *and in Jerusalem, speaking of the coming Day of the Lord, the biblical prophets in fact spoke of the Return of Nibiru.*

Ever since it rendered in the opening chapter of *Genesis* an abbreviated version of the Sumerian Epic of Creation, the Bible recognized the existence of Nibiru and its periodic return to Earth's vicinity, and treated it as another—in this case, celestial—manifestation of Yahweh as a Universal God. The Psalms and the Book of Job spoke of the unseen Celestial Lord that "in the heights of heaven marked out a circuit." They recalled this Celestial Lord's first appearance—when he collided with Tiamat (called in the Bible *Tehom* and nicknamed *Rahab* or *Rabah*, the Haughty One), smote her, created the heavens and "the Hammered Bracelet" (the Asteroid Belt), and "suspended the Earth in the void"; they also recalled the time when that celestial Lord caused the Deluge.

The arrival of Nibiru and the celestial collision, leading to Nibiru's great orbital circuit, were celebrated in the majestic Psalm 19:

> *The heavens bespeak the glory of the Lord;*
> *The Hammered Bracelet proclaims his handiwork . . .*
> *He comes forth as a groom from the canopy;*
> *Like an athlete he rejoices to run the course.*
> *From the end of the heavens he emanates,*
> *and his circuit is to their end.*

It was the nearing of the Celestial Lord at the time of the Deluge that was held to be the forerunner of what will happen next time the celestial Lord will return (Psalm 77: 6, 17–19):

> *I shall recall the Lord's deeds,*
> *remember thine wonders in antiquity . . .*
> *The waters saw thee, O Lord, and shuddered.*
> *Thine splitting sparks went forth,*
> *lightnings lit up the world.*
> *The sound of thine thunder was rolling,*
> *the Earth was agitated and it quaked.*

The Prophets considered those earlier phenomena as the guide for what to expect. They expected the Day of the Lord (to quote the Prophet Joel) to be a day when "the Earth shall be agitated, Sun and Moon shall be darkened, and the stars shall withhold their shining . . . A day that is great and terrifying."

The Prophets brought the word of Yahweh to Israel and all nations over a period of about three centuries. The earliest of the fifteen Literary Prophets was Amos; he began to be God's spokesman ("*Nabih*") circa **760 B.C.E.** His prophecies covered three periods or phases: he predicted the Assyrian assaults in the near future, a coming Day of Judgment, and an Endtime of peace and plenty. Speaking in the name of "the Lord Yahweh who reveals His secrets to the Prophets," he described the Day of the Lord as a day when "the Sun shall set at noon and the Earth shall darken in the midst of daytime." Addressing those who worship the "planets and star of their gods," he compared the coming Day to the events of the Deluge, when "the day darkened as night, and the waters of the seas poured upon the earth;" and he warned those worshippers with a rhetorical question (Amos 5: 18):

> *Woe unto you that desire the Day of the Lord!*
> *To what end is it for you?*
> *For the day of the Lord is darkness and no light.*

A half-century later, the Prophet Isaiah linked the prophecies of the "Day of the Lord" to a specific geographical site, to the "Mount of the Appointed Time," the place "on the northern slopes," and had this to say to the king who had set himself up on it: "Behold, the Day of the Lord cometh with pitiless fury and wrath, to lay the earth desolate and destroy the sinners upon it." He, too, compared what is about to happen to the Deluge, recalling the time when the "Lord came as a destroying tempest of mighty waves," and described (Isaiah 13: 10,13) the coming Day as a celestial occurrence that will affect the Earth:

> *The stars of heaven and its constellations*
> *shall not give their light;*
> *the Sun shall be darkened at its rising*
> *and the Moon shall not shine its light . . .*
> *The heavens shall be agitated*
> *and the Earth in its place will be shaken;*

When the Lord of Hosts shall be crossing
on the day of his wrath.

Most noticeable in this prophecy is the identification of the Day of the Lord as the time when "the Lord of Hosts"—the celestial, the planetary lord—*"shall be crossing."* This is the very language used in *Enuma elish* when it describes how the invader that battled Tiamat came to be called NIBIRU: *"Crossing* shall be its name!"

Following Isaiah, the Prophet Hosea also foresaw the Day of the Lord as a day when Heaven and Earth shall "respond" to each other—a day of celestial phenomena resonating on Earth.

As we continue to examine the prophecies chronologically, we find that in the seventh century B.C.E. the prophetic pronouncements became more urgent and more explicit: the Day of the Lord shall be a Day of Judgment upon the nations, Israel included, but primarily upon Assyria for what it has done and upon Babylon for what it will do, and *the Day is approaching, it is near*—

> *The great Day of the Lord is approaching—*
> *It is near!*
> *The sound of the Lord's Day hasteth greatly.*
> *A day of wrath is that day,*
> *a day of trouble and distress,*
> *a day of calamity and desolation,*
> *a day of darkness and deep gloom,*
> *a day of clouds and thick mist.*

> ZEPHANIA, I: 14–15

Just before 600 B.C.E. the Prophet Habakuk prayed to the "God who *in the nearing years is coming*," and who shall show mercy in spite of His wrath. Habakuk described the expected celestial Lord as a *radiant planet*—the very manner in which Nibiru was depicted in Sumer & Akkad. It shall appear, the Prophet said, from the southern skies:

The Lord from the south shall come . . .
Covered are the heavens with his halo,
His splendor fills the Earth.
His rays shine forth
from where his power is concealed.
The Word goes before him,
sparks emanate from below.
He pauses to measure the Earth;
He is seen, and the nations tremble.

HABAKKUK 3: 3–6

The prophecies' urgency increased as the sixth century B.C.E. began. ***"The Day of the Lord is at hand!"*** the Prophet Joel announced; ***"The Day of the Lord is near!"*** the Prophet Obadiah declared. Circa 570 B.C.E. the Prophet Ezekiel was given the following urgent divine message (Ezekiel 30: 2–3):

Son of Man, prophesy and say:
Thus sayeth the Lord God:
Howl and bewail for the Day!
For the Day is near—
the Day of the Lord is near!

Ezekiel was then away from Jerusalem, having been taken into exile with other Judean leaders by the Babylonian king Nebuchadnezzar. The place of exile, where Ezekiel's prophecies and famed vision of the Celestial Chariot took place, was on the banks of the Khabur River, in the region of **Harran**.

The location was not a chance one, for **the concluding saga of the Day of the Lord—and of Assyria and Babylon—was to be played out where Abraham's journey began.**

Chapter XII

DARKNESS AT NOON

W hile the Hebrew Prophets predicted Darkness at Noon, what were the "other nations" expecting as they awaited the Return of Nibiru?

To judge by their written records and engraved images, they were expecting the resolution of the gods' conflicts, benevolent times for mankind, and a great theophany. *They were in, as we shall see, for an immense surprise.*

In anticipation of the great event, the cadres of priests observing the skies in Nineveh and Babylon were mobilized to note celestial phenomena and interpret their omens. The phenomena were meticulously recorded and reported to the kings. Archaeologists have found in the remains of royal and temple libraries tablets with those records and reports that in many instances were arranged according to subject or the planet they were observing. A well-known collection in which some seventy tablets were combined—in antiquity—was a series titled *Enuma Anu Enlil*; it reported observations of planets, stars, and constellations classified according to the celestial Way of Anu and Way of Enlil—encompassing the skies from 30 degrees south all the way to zenith in the north (see Fig. 53).

At first the observations were interpreted by comparing the

phenomena to astronomical records from Sumerian times. Though written in Akkadian (Babylon's and Assyria's language), the observational reports extensively used Sumerian terminology and mathematics and sometimes included a scribal note that they were translated from earlier Sumerian tablets. Such tablets served as "astronomers' manuals," telling them from past experience what a phenomenon's oracular meaning was:

> When the Moon in its calculated time is not seen:
> There will be an invasion of a mighty city.

> When a comet reaches the path of the Sun:
> Field-flow will be diminished,
> an uproar will happen twice.

> When Jupiter goes with Venus:
> The prayers of the land will reach the gods.

As time went on, the reports were increasingly of observations accompanied by the omen-priests' own interpretations: "In the night Saturn came near to the Moon. Saturn is a planet of the Sun. This is the meaning: It is favorable to the king." The noticeable change included the paying of particular attention to eclipses; a tablet (now in the British Museum), listing computerlike columns of numbers, served to predict lunar eclipses fifty years in advance.

Modern studies have concluded that the change to the new style of topical astronomy took place in the eighth century B.C.E. when, after a period of mayhem and royal upheavals in Babylon and Assyria, the two lands' fates were placed in new and strong royal hands: Tiglath-Pileser III (745–727 B.C.E.) in Assyria and Nabunassar (747–734 B.C.E.) in Babylonia.

Nabunassar ("By Nabu protected") was hailed, already in antiquity, as an innovator and powerhouse in the field of astronomy. One of his first actions was to repair and restore the temple of

Shamash in Sippar, the Sun-god's "cult center" in ancient Sumer. He also built a new observatory in Babylon, updated the calendar (a heritage from Nippur), and instituted daily reporting to the king of the celestial phenomena and their meaning. It was primarily due to those measures that a wealth of astronomical data, shedding light on subsequent events, has come to light.

Tiglath-Pileser III was also active, in his own ways. His annals describe constant military campaigns and boast of captured cities, brutal executions of local kings and nobility, and mass exiles. His role, and those of his successors Shalmaneser V and Sargon II, in the demise of Israel and the exile of its people (the Ten Lost Tribes), and then the attempt by Sennacherib to seize Jerusalem, were described in the previous chapter. Closer to home, those Assyrian kings were busy annexing Babylonia by "taking the hands of Marduk." The next Assyrian king, Esarhaddon (680–669 B.C.E.), announced that "both Ashur and Marduk gave me wisdom," swore oaths in the name of Marduk and Nabu, and started to rebuild the Esagil temple in Babylon.

In history books, Esarhaddon is mainly remembered for his successful invasion of Egypt (675–669 B.C.E.). The invasion's purpose, as far as it could be ascertained, was to stop Egyptian attempts to "meddle in Canaan" and dominate Jerusalem. Noteworthy, in the light of subsequent events, was the route he chose: instead of going the shortest way, to the southwest, he made a considerable detour and went northward, to **Harran**. There, in the olden temple of the god Sin, Esarhaddon sought that god's blessing to embark on the conquest; and Sin, leaning on a staff and accompanied by Nusku (the Divine Messenger of the gods), gave his approval.

Esarhaddon then did turn southward, sweeping mightily through the lands of the eastern Mediterranean to reach Egypt. Significantly, he detoured away from the prize that Sennacherib failed to seize—Jerusalem. Significantly, too, that invasion of Egypt and the detour away from Jerusalem—as well as Assyria's own eventual fate—had been prophesied by Isaiah decades earlier (10: 24–32).

Busy geopolitically as Esarhaddon was, he did not neglect the astronomical requirements of those times. With guidance from the gods Shamash and Adad, he erected in Ashur (the city, Assyria's cult center) a "House of Wisdom"—an observatory—and depicted the complete twelve-member solar system, including Nibiru, on his monuments **(Fig. 89)**. Leading to a more lavish sacred precinct was a new monumental gate, built—according to cylinder seal

FIGURE 89

FIGURE 90

depictions—to emulate Anu's gateway on Nibiru **(Fig. 90)**. It is a clue to what the Return expectations in Assyria were.

All those religious-political moves suggest that the Assyrians made sure to "touch all the bases" as far as the gods were concerned. And so, by the seventh century B.C.E., Assyria was ready for the anticipated Return of the planet of the gods. Discovered texts—including letters to the kings by their chief astronomers—reveal anticipation of an idyllic, utopian time:

> *When Nibiru will culminate . . .*
> *The lands will dwell securely,*
> *Hostile kings will be at peace;*
> *The gods will receive prayers*
> *and hear supplications.*
>
> *When the Planet of the Throne of Heaven*
> *will grow brighter,*
> *there will be floods and rains.*
>
> *When Nibiru attains its perigee,*
> *the gods will give peace.*
> *Troubles will be cleared up,*
> *complications will be unravelled.*

Clearly, the expectation was of a planet that will appear, rise in the skies, grow brighter, and at its perigee, at the Crossing, become NIBIRU (the Cross Planet). And as the gateway and other construction indicated, with the returning planet *a repeat of the previous visit to Earth by Anu was expected.* It was now up to the astronomer-priests to watch the heavens for that planetary appearance; but where were they to look in the celestial expanse, and how would they recognize the planet when still in the distant skies?

The next Assyrian king, Ashurbanipal (668–630 B.C.E.), came up with a solution.

Historians consider Ashurbanipal to have been the most scholarly of the Assyrian kings, for he had learnt other languages besides Akkadian, including Sumerian, and claimed that he could even read "writings from before the Flood." He also boasted that he "learnt the secret signs of Heaven and Earth . . . and studied the heavens with the masters of divination."

Some modern researchers also consider him to have been "The First Archaeologist," for he systematically collected tablets from sites that were already ancient in his time—like Nippur, Uruk, and Sippar in what used to be Sumer. He also sent specialized teams to sort out and loot such tablets from the capitals that the Assyrians overran. The tablets ended up in a famed library where teams of scribes studied, translated, and copied chosen texts from the previous millennia. (A visitor to the Museum of the Ancient Near East in Istanbul can see a display of such tablets, neatly arranged on the original shelves, with each shelf headed by a "catalog tablet" that lists all the texts on that shelf.)

While the subjects in the accumulated tablets covered a wide range, what was found indicates that particular attention was given to celestial information. Among the purely astronomical texts there were tablets that belonged to a series titled "**The day of Bel**"—the *Day of the Lord!* In addition, epic tales and histories pertaining to the gods' comings and goings were deemed important, especially if they shed light on Nibiru's passages. *Enuma elish*—the Epic of Creation that told how an invading planet joined the solar system to become Nibiru—was copied, translated, and recopied; so were writings dealing with the Great Flood, such as the *Atra-Hasis Epic* and the *Epic of Gilgamesh*. While they all seem to legitimately be part of accumulating knowledge in a royal library, it so happens that *they all dealt with instances of Nibiru's appearances in the past*—and thus with its next nearing.

Among the purely astronomical texts translated and, undoubtedly, carefully studied, were guidelines for observing Nibiru's arrival and for recognizing it on its appearance. A Babylonian text that retained the original Sumerian terminology stated:

Planet of the god Marduk:
Upon its appearance SHUL.PA.E;
Rising thirty degrees, SAG.ME.NIG;
When it stands in the middle of the sky: NIBIRU.

While the first-named planet (SHUL.PA.E) is deemed to be Jupiter (but could be Saturn), the next one's name (SAG.ME.NIG) could just be a variant for Jupiter, but is considered by some to be Mercury (★). A similar text from Nippur, rendering the Sumerian planetary names as UMUN.PA.UD.DU and SAG.ME.GAR, suggested that the arrival of Nibiru will be "announced" by the planet Saturn, and after rising 30 degrees will be near Jupiter. Other texts (e.g., a tablet known as K.3124) state that after passing SHUL.PA.E and SAG.ME.GAR—which I believe mean Saturn and Jupiter—"Planet Marduk" will "enter the Sun" (i.e., reach Perigee, closest to the Sun) and "become Nibiru."

★The extensive astronomical data that have been found attracted, already in the 19th and early in the 20th centuries, the time, attention, and patience of scholarly giants who brilliantly combined "Assyriology" with knowledge of astronomy. The very first book of The Earth Chronicles, *The 12th Planet*, covered and used the work and achievements of the likes of Franz Kugler, Ernst Weidner, Erich Ebeling, Herman Hilprecht, Alfred Jeremias, Morris Jastrow, Albert Schott, and Th. G. Pinches, among others. Their task was complicated by the fact that the same *kakkabu* (any celestial body, including planets, fixed stars, and constellations) could have more than one name. I also pointed out right then and there the most basic failing of their work: they all assumed that the Sumerians and other ancient peoples had no way of knowing ("with the naked eye") about planets beyond Saturn. The result was that whenever a planet was named other than the accepted names for the "seven known *kakkabani*"— Sun, Moon, Mercury, Venus, Mars, Jupiter, Saturn—it was assumed to just be yet another name for one of those "known seven." The principal victim of that erroneous stance was Nibiru; whenever it or its Babylonian equivalent "planet Marduk" was listed, it was assumed to be another name for Jupiter or Mars or (in some extreme views) even for Mercury. Incredibly, modern establishment astronomers continue to base their work on that "only seven" assumption—in spite of the vast contrary evidence that shows that the Sumerians knew the true shape and composition of our solar system, starting with the naming of the outer planets in *Enuma elish*, or the 4,500-year-old depiction of the complete twelve-member solar system, with the Sun in the center, on cylinder seal VA/243 in the Berlin Museum **(Fig. 91)**, or the depiction of twelve planetary symbols on Assyrian and Babylonian monuments, etc.

FIGURE 91

Other texts provide clearer clues in regard to Nibiru's path, as well as to the time frame for its appearance:

> *From the station of Jupiter,*
> *the planet passes toward the west.*
>
> *From the station of Jupiter*
> *the planet increases its brilliance,*
> *and in the zodiac of Cancer will become Nibiru.*
>
> *The great planet:*
> *At his appearance: Dark red.*
> *The heaven he divides in half*
> *as it stands in Nibiru.*

Taken together, the astronomical texts from the time of Ashurbanipal described a planet appearing from the solar system's edge, rising and becoming visible when it reaches Jupiter (or even Saturn before that), and then curving down toward the ecliptic. At its perigee, when it is closest to the Sun (and thus to Earth), the planet—at the Crossing—becomes Nibiru **"in the zodiac of Cancer."** That, as the enclosed schematic (and not to scale) diagram shows, could happen only when sunrise on the day of the Spring Equinox took place in the Age of the Ram—during the zodiacal age of Aries **(Fig. 92)**.

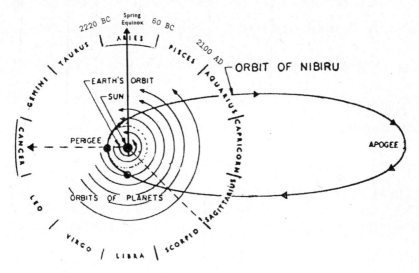

FIGURE 92

Such clues regarding the orbital path of the Celestial Lord and its reappearance, sometimes using the constellations as a celestial map, are also found in biblical passages, thereby revealing knowledge that must have been internationally available:

"In Jupiter will thy face be seen," states Psalm 17. "The Lord from the south shall come . . . his shining splendor will beam as light," predicted the prophet Habakkuk (Chapter 2). "Alone he stretches out the heavens and treads upon the highest Deep; he arrives at the Great Bear, Sirius and Orion, and the constellations of the south," the Book of Job (Chapter 9) stated; and the Prophet Amos (5: 9) foresaw the Celestial Lord "smiling his face upon Taurus and Aries, from Taurus to Sagittarius he shall go." These verses described a planet that spans the highest heavens and, *orbiting clockwise*—"retrograde," astronomers say—arrives via the southern constellations. It is a trajectory, on a vaster scale, akin to that of Halley's comet (see Fig. 78).

A telling clue in regard to Ashurbanipal's expectations was the meticulous rendering into Akkadian of Sumerian descriptions of the ceremonies attending the state visit of Anu and Antu to Earth circa 4000 B.C.E. The sections dealing with their stay in Uruk described how, at evetime, an observer was stationed "on the topmost stage of the tower" to watch for and announce the appearance of the planets one after the other, until the "Planet of the Great Anu of Heaven" came into view, whereupon all the gods assembled to welcome the divine couple recited the composition "To the one who grows bright, the heavenly planet of the god Anu" and sang the hymn "The Creator's image has arisen." The long texts then described the ceremonial meals, the retreat to the nighttime chambers, the processions the next day, and so on.

One can reasonably conclude that Ashurbanipal was engaged in collecting, collating, translating, and studying all the earlier texts that could (a) provide guidance to the astronomer-priests for detecting, at the first possible moment, the returning Nibiru and (b) inform the king about the procedures for what to do next. Calling the planet "Planet of the Heavenly Throne" is an important clue to the royal expectations, as were the depictions on palace walls, in magnificent reliefs, of Assyrian kings *greeting the god in the Winged Disc* as it hovered above the Tree of Life (as in Fig. 87).

It was important to be informed of the planet's appearance as soon as possible in order to be able to *prepare the proper reception for the arrival of the great god depicted within it*—Anu himself?—and be blessed with long, perhaps even eternal, Life.

But that was not destined to be.

Soon after Ashurbanipal's death, rebellions broke out throughout the Assyrian empire. His sons' hold on Egypt, Babylonia, and Elam disintegrated. Newcomers from afar appeared on the borders of the Assyrian empire—"hordes" from the north, Medes from the east. Everywhere, local kings seized control and declared independence. Of particular importance—immediate and for future events—was Babylon's "decoupling" of the dual kingship with Assyria. As part of the New Year festival in 626 B.C.E. a Babylonian general whose name—Nabupolassar ("Nabu his son protects")—implied that he claimed to be a son of the god Nabu, was enthroned as king of an independent Babylonia. A tablet described the start of his investiture ceremony thus: "The princes of the land were assembled; they blessed Nabupolassar; opening their fists, they declared him sovereign; Marduk in the assembly of the gods gave the Standard of Power to Nabupolassar."

The resentment of Assyria's brutal rule was so great that Nabupolassar of Babylon soon found allies for military action against Assyria. A principal and freshly vigorous ally was the Medes (precursors of the Persians), who had experienced Assyrian incursions and brutality. While Babylonian troops were advancing into Assyria from the south, the Medes attacked from the east, and in 614 B.C.E.—as had been prophesied by the Hebrew Prophets!—captured and burned down Assyria's religious capital, Ashur. The turn of Nineveh, the royal capital, came next. *By 612 B.C.E. the great Assyria was in shambles.* Assyria—the land of the "First Archaeologist"—itself became a land of archaeological sites.

How could that happen to the land whose very name meant "Land of the god Ashur"? The only explanation at the time was that the gods withdrew their protection from that land; in fact, we shall show, there was much more to it: **the gods themselves withdrew—from that land and from Earth.**

And then the most astounding and final chapter of the Return Saga, in which **Harran** was to play a key role, began to unfold.

The amazing chain of events after the demise of Assyria began with the escape to **Harran** of members of Assyria's royal family.

Seeking there the protection of the god Sin, the escapees rallied the remnants of the Assyrian army and proclaimed one of the royal refugees as "King of Assyria"; but the god, whose city Harran has been since days of yore, did not respond. *In 610 B.C.E. Babylonian troops captured Harran and put an end to the Assyrians' lingering hopes.*

The contest for the mantle of successorship to the heritage of Sumer and Akkad was over; it was now worn solely, and with divine blessing, by the king in Babylon. Again, Babylon ruled the lands that were once the hallowed "Sumer & Akkad"—so much so that in many texts from that time, Nabupolassar was given the title "King of Akkad." He used that authority to extend the celestial observations to the erstwhile Sumerian cities of Nippur and Uruk, and some of the key observational texts from the subsequent crucial years come from there.

It was in that same fateful year, 610 B.C.E.—a memorable year of astounding events, as we shall see—that a reinvigorated Egypt also placed on its throne an assertive strongman named Necho. Just one year later one of the least understod—by historians, that is—geopolitical moves then took place. The Egyptians, who used to be on the same side as the Babylonians in opposition to Assyrian rule, emerged from Egypt and, rushing northward, overran territories and sacred sites that the Babylonians considered theirs. The Egyptian advance, all the way north to Carchemish, put them within striking distance of Harran; it also placed in Egyptian hands the two space-related sites, in Lebanon and in Judea.

The surprised Babylonians were not going to let it stand. The aging Nabupolassar entrusted the task of recapturing the vital places to his son Nebuchadnezzar, who had already distinguished himself on the battlefields. In June 605 B.C.E., at Carchemish, the Babylonians crushed the Egyptian army, liberated "the sacred forest in Lebanon which Nabu and Marduk desired," and chased the fleeing Egyptians all the way to the Sinai Peninsula. Nebuchadnezzar stopped the pursuit only on news from Babylon that his father had died. He rushed back, and was proclaimed King of Babylon that same year.

Historians find no explanation for the sudden Egyptian thrust and the ferocity of the Babylonian reaction. To us it is evident that at the core of the events was the expectation of the Return. Indeed, it seems that in that year 605 B.C.E. the Return was deemed to be imminent, perhaps even overdue; for it was in that very same year that the Prophet Habakkuk began to prophecy in the name of Yahweh, in Jerusalem.

Uncannily foretelling the future of Babylon and other nations, the Prophet asked Yahweh when the Day of the Lord—a day of judgment upon the nations, Babylon included—would come, and Yahweh responded, saying:

> *Write down the prophecy,*
> *explain it clearly on the tablets,*
> *so that it may be quickly read:*
> *For the vision there is a set time;*
> *In the end it shall come, without fail!*
> *Though it may tarry, wait for it;*
> *For it will surely come—*
> *For its appointed time it will not be delayed.*
>
> HABAKKUK 2: 2–3

(The "appointed time," as we shall see, arrived precisely fifty years thereafter.)

The forty-three years of Nebuchadnezzar's reign (605–562 B.C.E.) are considered a period of a dominant "Neo-Babylonian" empire, a period marked by decisive actions and fast moves, for there was no time to lose—the nearing Return was now Babylon's prize!

To prepare Babylon for the expected Return, massive renovation and construction works were quickly undertaken. Their focal point was the sacred precinct, where the Esagil temple of Marduk (now simply called *Bel/Ba'al*, "The Lord") was renovated and rebuilt, its seven-stage ziggurat readied for viewing from it the

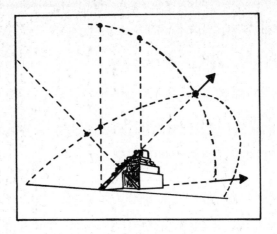

FIGURE 93

starry skies **(Fig. 93)**—just as had been done in Uruk when Anu had visited circa 4000 B.C.E. A new Processional Way leading to the sacred precinct through a massive new gate was erected; their walls were decorated and covered from top to bottom with artful glazed bricks that astound to this day, for the site's modern excavators have removed and put the Processional Way and the Gate back together at the Vorderasiatiches Museum in Berlin. Babylon, Marduk's Eternal City, was readied to welcome the Return.

"I have made the city of Babylon to be the foremost among all the countries and every habitation; its name I elevated to be the most praised of all the sacred cities," Nebuchadnezzar wrote in his inscriptions. The expectation, it seems, was that the arriving god of the Winged Disk would come down at the Landing Place in Lebanon, then consummate the Return by entering Babylon through the new

FIGURE 94

marvelous Processional Way and imposing gate **(Fig. 94)**—a gate named "Ishtar" (alias IN.ANNA), who had been "Anu's beloved" in Uruk—another clue regarding whose Return was expected.

Accompanying these expectations was Babylon's role as the new Navel of the Earth—inheriting the prediluvial status of Nippur as the DUR.AN.KI, the "Bond Heaven-Earth." That this was now Babylon's function was expressed by giving the ziggurat's foundation platform the *Sumerian* name E.TEMEN.AN.KI ("Temple of the Foundation for Heaven-Earth"), stressing Babylon's role as the new "Navel of the Earth"—a role clearly depicted on the Babylonian "Map of the World" (see Fig. 10).

This was terminology that echoed the description of Jerusalem, with its Foundation Stone, serving as a link between Earth and Heaven!

But if that was what Nebuchadnezzar envisioned, then Babylon had to replace the existing post-Diluvial space link—Jerusalem.

Having taken over Nippur's prediluvial role to serve as Mission Control Center after the Deluge, Jerusalem was located at the center of concentric distances to the other space-related sites (see Fig. 3). Calling it the "Navel of the Earth" (38: 12), the Prophet Ezekiel announced that Jerusalem has been chosen for this role by God himself:

> *Thus has said the Lord Yahweh:*
> *This is Jerusalem;*
> *In the midst of the nations I placed her,*
> *and all the lands are in a circle*
> *round about her.*
>
> EZEKIEL 5: 5

Determined to usurp that role for Babylon, Nebuchadnezzar led his troops to the elusive prize and in 598 B.C.E. captured Jerusalem. This time, as the Prophet Jeremiah had warned, Nebuchadnezzar was carrying out God's anger at Jerusalem's people, for they had taken up the worship of the celestial gods: "Ba'al, the Sun and the Moon, and the constellations" (II Kings 23: 5)—*a list that clearly included Marduk as a celestial entity!*

Starving Jerusalem's people by a siege that lasted three years, Nebuchadnezzar managed to subdue the city and took the Judean king Jehoyachin captive to Babylon. Taken into exile were also Judea's nobility and learned elite—among them the Prophet Ezekiel—and thousands of its soldiers and craftsmen; they were made to reside by the banks of the Khabur River, near Harran, their ancestral home.

The city itself and the Temple were left intact this time, but eleven years later, in 587 B.C.E., the Babylonians returned in force. Acting this time, according to the Bible, on their own volition, the Babylonians put the torch to the Temple that Solomon built. In his inscriptions Nebuchadnezzar offered no explanation other than the usual one—to carry out the wishes of and to please "my gods Nabu and Marduk"; but as we shall soon show, the real reason was a simple one: a belief that Yahweh had departed and was gone.

The destruction of the Temple was a shocking and evil deed for which Babylon and its king—previously deemed by the Prophets to have been Yahweh's "rod of wrath"—were to be severely punished: "The vengeance of Yahweh our God, **vengeance for His Temple**," shall be meted out to Babylon, announced the Prophet Jeremiah (50: 28). Predicting the fall of mighty Babylon and its destruction by invaders from the north—events that came true just a few decades later—Jeremiah also proclaimed the fate of the gods whom Nebuchadnezzar had invoked:

> *Declare among the nations and proclaim,*
> *Raise the sign, announce, do not conceal,*
> *Say: Captured is Babylon!*
> *Withered is* Bel, *confounded is* Marduk!
>
> JEREMIAH 50: 2

Divine punishment upon Nebuchadnezzar himself was commensurate with the sacrilege. Crazed, according to traditional sources, by a bug that entered his brain through his nose, Nebuchadnezzar died in agony in 562 B.C.E.

Neither Nebuchadnezzar nor his three bloodline successors (who were murdered or otherwise disposed of in short shrift) lived to see an arrival of Anu at the gates of Babylon. In fact, **such an arrival never took place, even though Nibiru did return.**

*It is a fact that astronomical tablets from that very time record
actual observations of Nibiru, alias "Planet of Marduk."* Some
were reported as omina, for example, a tablet catalogued K.8688
that informed the king that if Venus shall be seen "in front of"
(i.e., rising ahead of) Nibiru, the crops will fail, but if Venus shall
rise "behind" (i.e., after) Nibiru, "the crop of the land will suc-
ceed." Of greater interest to us are a group of "Late Babylonian"
tablets found in Uruk; they rendered the data in twelve monthly
zodiacal columns and combined the texts with pictorial depictions. In one of these tablets (VA 7851, **Fig. 95**), the Planet of
Marduk, shown between the Aries ram symbol on one side and
the seven symbol for Earth on the other side, depicts Marduk

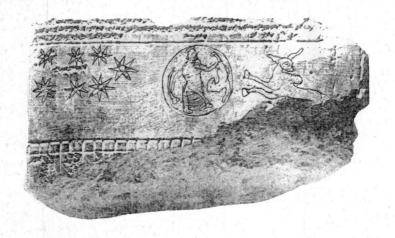

FIGURE 95

within the planet. Another example is tablet VAT 7847; it names
an actual observation, **in the constellation of Aries,** as the **"Day
when the gate of the great lord Marduk was opened"**—when Nibiru
had appeared into view; and then has an entry—"*Day of the Lord
Marduk*"—as the planet moved on and was seen in Aquarius.

Even more telling of the coming into view of the planet "Marduk" from the southern skies and its fast becoming "Nibiru" in the central celestial band, were yet another class of tablets, this time circular. Representing "an advance backward" to the Sumerian astronomical tenets, the tablets divided the celestial sphere into the three Ways (Way of Enlil for the northern skies, of Ea for the southern, and of Anu in the center). The twelve zodiacal-calendrical segments were then superimposed on the three Ways, as shown by the discovered fragments **(Fig. 96)**; explanatory texts were written on the back sides of those circular tablets.

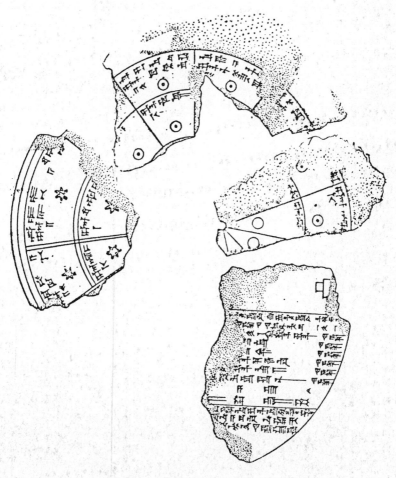

FIGURE 96

In A.D. 1900, addressing a meeting of the Royal Asiatic Society in London, England, Theophilius G. Pinches caused a sensation when he announced that he had succeeded in piecing together a complete "astrolabe" ("Taker of Stars"), as he called the tablet. He showed it to be a circular disc divided into three concentric sections and, like a pie, into twelve segments, resulting in a field of thirty-six portions. Each of the thirty-six portions contained a name with a small circle below it, indicating it was a celestial body, and a number. Each portion also bore a month's name, so Pinches numbered them from I to XII, starting with Nissan **(Fig. 97)**.

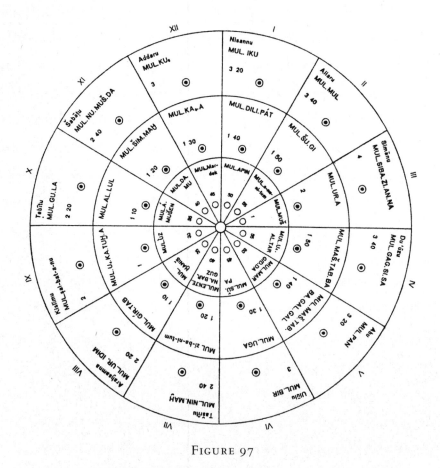

FIGURE 97

The presentation caused an understandable sensation, for here was a Babylonian sky map, divided into the three Ways of Enlil, Anu, and Ea/Enki, showing which planets, stars, and constellations were observed where at each month during the year. The debate over the identity of the celestial bodies (at the root of which lurks the notion of "nothing beyond Saturn") and the meaning of the numbers has yet to end. Also unresolved is the issue of dating— in what year was the astrolabe made, and if it was a copy of an earlier tablet, what was the time shown? Dating opinions ranged from before the twelfth century to the third century B.C.E.; most agreed, however, that the astrolabe belonged to the era of Nebuchadnezzar or his successor Nabuna'id.

The astrolabe presented by Pinches was identified in the ensuing debates as "P," but has been later renamed "Astrolabe A" because another one has since been pieced together and is known as "Astrolabe B."

Though the two astrolabes at first glance look identical, they are different—and for our analysis, the key difference is that in "B" the planet identified as *mul Neberu deity Marduk*—"Planet Nibiru of the god Marduk"—is shown in the Way of Anu, the central-ecliptic band **(Fig. 98)**, whereas in "A" the planet identified as *mul Marduk*—the "Planet of Marduk"—is shown in the Way of Enlil, in the northern skies **(Fig. 99)**.

The change in name and position is absolutely correct if the two astrolabes depict a *moving planet*—"Marduk" as it was called by the Babylonians—that, after having come into view high in the northern skies (as in "A"), curves down to cross the ecliptic and becomes NIBIRU ("Crossing") when it crosses the ecliptic *in the Way of Anu* (as in "B"). The two-stage documentation by the two astrolabes depicts precisely what we have been asserting all along!

The texts (known as KAV 218, columns B and C) accompanying the circular depictions remove any shadow of doubt regarding the Marduk/Nibiru identity:

[Month] Adar:
Planet Marduk in the Way of Anu:
The radiant Kakkabu *which rises in the south*
after the gods of the night finished their tasks,
and divides the heavens.
This kakkabu *is Nibiru = god Marduk.*

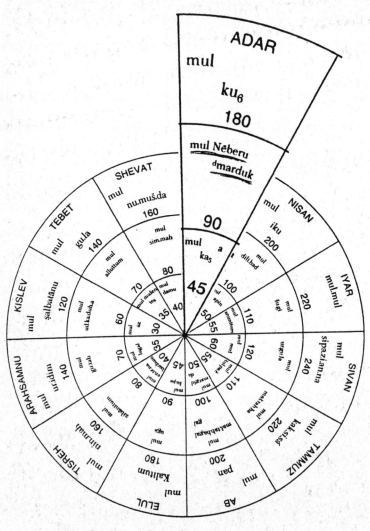

FIGURE 98

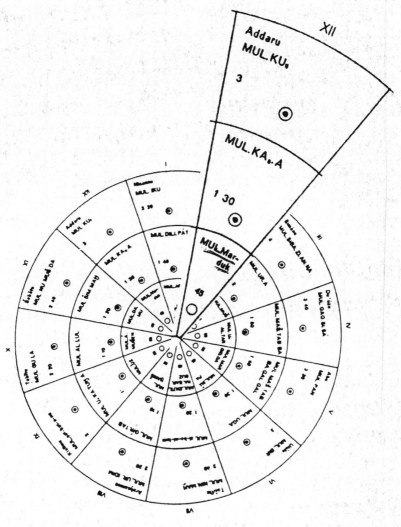

FIGURE 99

While we can be certain—for reasons soon to be given—that
the observations in all those "Late Babylonian" tablets could not
have taken place earlier than 610 B.C.E., we can also be sure that
they did not take place after 555 B.C.E., for that was the date when
one called Nabuna'id became the last king of Babylonia; and his
claim to legitimacy was that his kingship was celestially confirmed

because *"the planet of Marduk, high in the sky, had called me by my name."* Making that claim, he also stated that in a nighttime vision he had seen *"the Great Star and the Moon."* Based on the Kepler formulas for planetary orbits around the Sun, the whole period of Marduk/Nibiru's visibility from Mesopotamia lasted just a short few years; hence, the visibility claimed by Nabuna'id places the planet's Return in the years immediately preceding 555 B.C.E.

So when was the precise time of the Return? There is one more aspect involved in resolving the puzzle: the prophecies of "Darkness at noon" on the Day of the Lord—a solar eclipse— **and such an eclipse did in fact occur, in 556 B.C.E.!**

Solar eclipses, though much rarer than lunar eclipses, are not uncommon; they happen when the Moon, passing in a certain way between Earth and the Sun, temporarily obscures the Sun. Only a small portion of solar eclipses are total. The extent, duration, and path of total darkness vary from passage to passage due to the ever-changing triple orbital dance between Sun, Earth, and Moon, plus Earth's daily revolution and its changing axial tilt.

As rare as solar eclipses are, the astronomical legacy of Mesopotamia included knowledge of the phenomenon, calling it *atalu shamshi*. Textual references suggest that not only the phenomenon but even its lunar involvement were part of the accumulated ancient knowledge. In fact, a solar eclipse whose path of totality passed over Assyria had occurred in 762 B.C.E. It was followed by one in 584 B.C.E. that was seen all across the Mediterranean lands, with totality over Greece. **But then, in 556 B.C.E., there occurred an extraordinary solar eclipse** *"not in an expected time."* If it was not due to the predictable motions of the Moon, **could it have been caused by an unusually close passage of Nibiru?**

Among astronomical tablets belonging to a series called "When Anu Is Planet of the Lord," one tablet (catalogued VACh.Shamash/ RM.2,38—**Fig. 100**), dealing with a solar eclipse, recorded thus the observed phenomenon (lines 19–20):

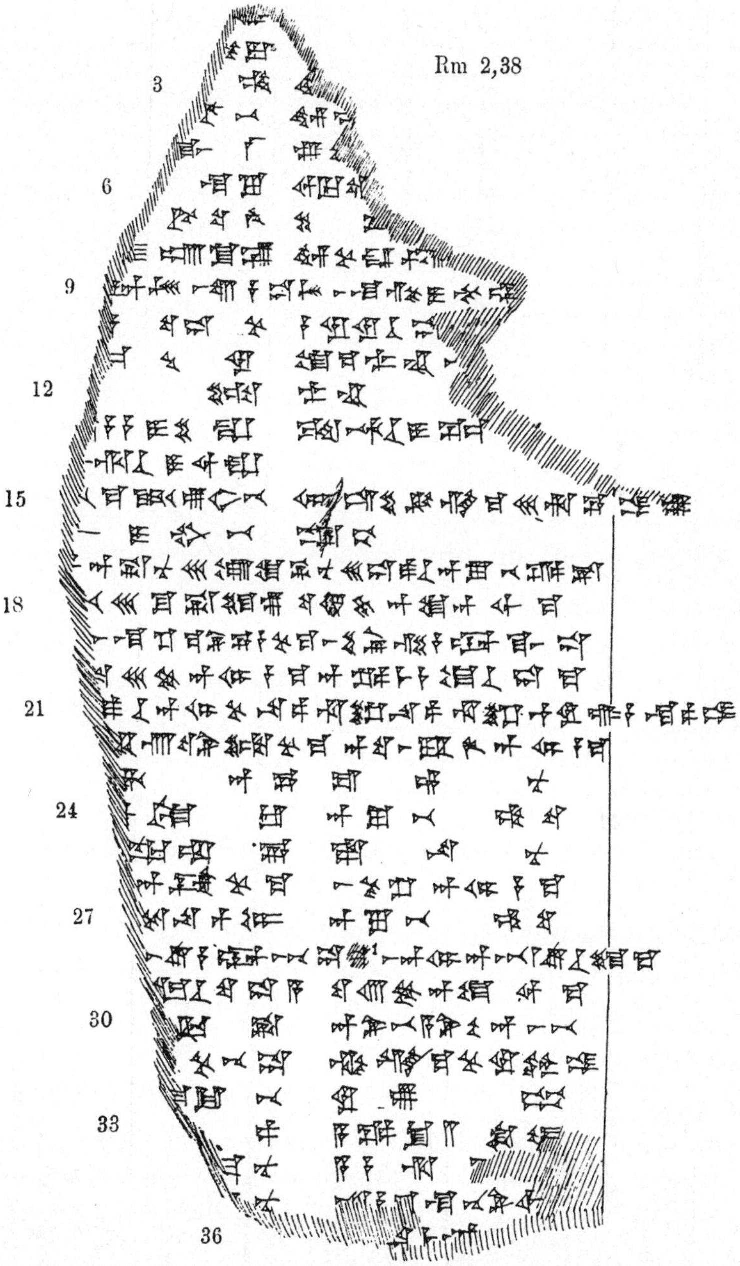

Rm 2,38

FIGURE 100

In the beginning the solar disc,
not in an expected time,
became darkened,
and **stood in the radiance of the Great Planet.**
On day 30 [of the month] was
the **eclipse of the Sun.**

What exactly do the words that the darkened Sun "stood in the radiance of the Great Planet" mean? Though the tablet itself does not provide a date for that eclipse, it is our suggestion that the particular wording, highlighted above, **strongly indicates that the unexpected and extraordinary solar eclipse was somehow caused by the return of Nibiru,** the "great radiating planet"; but whether the direct cause was the planet itself, or the effects of its "radiance" (gravitational or magnetic pull?) on the Moon, the texts do not explain.

Still, it is an astronomically historic fact that on a day equal to May 19, 556 B.C.E., a major total solar eclipse did occur. As shown by this map, prepared by NASA's Goddard Space Flight Center **(Fig. 101)**, the eclipse was a great and major one, seen over wide areas, and a unique aspect about it was that **the band of total darkness passed exactly over the district of Harran!**

This last fact is of the utmost significance for our conclusions—and it was even more so in those fateful years in the ancient world; for *right after that,* in 555 B.C.E., Nabuna'id was proclaimed king of Babylonia—not in Babylon, but in Harran. He was the last king of Babylon; after him, as Jeremiah had prophesied, Babylon followed the fate of Assyria.

It was in 556 B.C.E. that the prophesied Darkness at Noon came. It was just then that Nibiru returned; it was the prophesied DAY OF THE LORD.

And when the planet's Return did occur, neither Anu

nor any other of the expected gods showed up. Indeed, the opposite happened: the gods, the Anunnaki gods, took off and left the Earth.

Total Solar Eclipse of -0556 May 19

Geocentric Conjunction = 12:50:16.9 UT   J.D. = 1518118.034918
Greatest Eclipse = 12:44:22.5 UT   J.D. = 1518118.030815

Eclipse Magnitude = 1.02584   Gamma = 0.31810

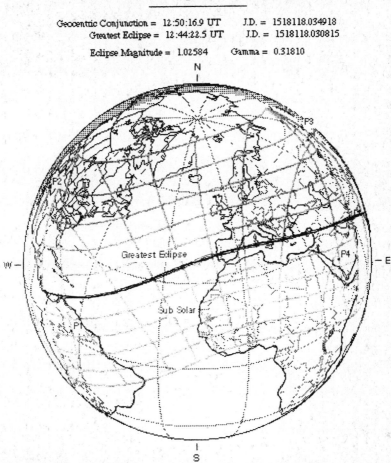

FIGURE 101

237

Chapter XIII

WHEN THE GODS LEFT EARTH

The departure of the Anunnaki gods from Earth was a drama-filled event replete with theophanies, phenomenal occurrences, divine uncertainties, and human quandary.

Incredibly, the Departure is neither surmised nor speculative; it is amply documented. The evidence comes to us from the Near East as well as from the Americas; and some of the most direct, and certainly the most dramatic, records of the ancient gods' departure from Earth come to us from **Harran**. The testimony is not hearsay; it consists of *eyewitness reports*, among them by the Prophet Ezekiel. The reports are included in the Bible, and they were inscribed on stone columns—texts dealing with miraculous events leading to the accession to the throne of Babylon's last king.

Harran nowadays—yes, it is still there, and I have visited it—is a sleepy town in eastern Turkey, just a few miles from the Syrian border. It is surrounded by crumbling walls from Islamic times, its inhabitants dwelling in beehive-shaped mud huts. The traditional well where Jacob met Rachel is still there among the sheep meadows outside the town, with the purest naturally cool water one can imagine.

But in earlier days Harran was a flourishing commercial, cultural, religious, and political center, so much so that even the Prophet Ezekiel (27: 24), who lived in the area with other exiles from Jerusalem, recalled her reputation as a trader in "blue clothes and broidered work, and in chests of rich apparel, bound with cords and made of cedar." It was a city that had been from Sumerian times on an "Ur away from Ur" cult center of the "Moon god" Nannar/Sin. Abraham's family ended up residing there because his father Terah was a *Tirhu*, an omen-priest, first in Nippur, then in Ur, and finally in Nannar/Sin's temple in Harran. After the demise of Sumer by the nuclear Evil Wind, Nannar and his spouse, Ningal, made their home and headquarters in Harran.

Though Nannar ("*Su-en,*" or *Sin* for short in Akkadian) was not Enlil's firstborn legal heir—that rank belonged to Ninurta—he was the firstborn of Enlil and his spouse Ninlil, a firstborn on Earth. Gods and men greatly adored Nannar/Sin and his spouse; the hymns in their honor in Sumer's glorious times, and the lamentations about the desolation of Sumer in general and Ur in particular, reveal the great love and admiration of the people for this divine couple. That many centuries later Esarhaddon went to consult with an aging Sin ("leaning on a staff") regarding the invasion of Egypt, and that the escaping Assyrian royals made a last stand in Harran, serve to indicate the continued important role played by Nannar/Sin and Harran to the very end.

It was in the ruins of the city's great Nannar/Sin temple, the E.HUL.HUL ("House of Double Joy"), that archaeologists discovered four stone columns ("stelae") that once stood in the temple, one at each corner of the main prayer hall. The inscriptions on the stelae revealed that two were erected by the temple's high priestess, Adda-Guppi, and two by her son Nabuna'id, the last king of Babylon.

With an evident sense of history and as a trained temple official, Adda-Guppi provided in her inscriptions precise dates for the astounding events that she had witnessesd. The dates, linked as was then customary to regnal years of known kings, could thus be—and have been—verified by modern scholars. It is thus certain that she was born in 649 B.C.E. and lived through the reigns of several Assyrian and Babylonian kings, passing on at the ripe old age of 104.

Here is what she wrote on her stela concerning the first of a series of amazing events:

> *It was in the sixteenth year of Nabupolassar,*
> *king of Babylon, when Sin, lord of the gods,*
> *became angry with his city and his temple*
> **and went up to heaven;**
> *and the city and the people in it went to ruin.*

The sixteenth year of Nabupolassar was 610 B.C.E.—a memorable year, the reader may recall, when Babylonian forces captured Harran from the remnants of the Assyrian royal family and army, and when a reinvigorated Egypt decided to seize the space-related sites. It was then, Adda-Guppi wrote, that an angered Sin, removing his protection (and himself) from the city, packed up **"and went up to heaven**!**"**

What followed in the captured city is accurately summed up: "And the city and its people went to ruin." While other survivors fled, Adda-Guppi stayed on. "Daily, without ceasing, by day and night, for months, for years," she kept vigil in the ruined temple. Mourning, she "forsook the dresses of fine wool, took off jewelry, wore neither silver nor gold, relinquished perfumes and sweet smelling oils." As a ghost roaming the abandoned shrine, "in a torn garment I was clothed; I came and went noiselessly," she wrote.

Then, in the desolate sacred precinct, she found a robe that had once belonged to Sin. To the despondent priestess, the find was an omen from the god: suddenly he had given her a physical presence of himself. She could not take her eyes off the sacred garb, not daring to touch it except by "taking hold of its hem." As if the god himself was there to hear her, she prostrated herself and "in prayer and humility" uttered a vow: "If you would return to your city, all the Black-Headed people would worship your divinity!"

"Black-Headed people" was a term by which the Sumerians used to describe themselves, and the employment of the term by the high priestess some 1,500 years after Sumer was no more was full of significance: she was telling the god that were he to come back, he would be restored to lordship as in the Days of Old, become again the lord god of a restored Sumer and Akkad. To achieve that, Adda-Guppi offered her god a deal: If he would return and then use his divine powers to make her son Nabuna'id the next imperial king, reigning over all the Babylonian and Assyrian domains, Nabuna'id would restore the temple of Sin not only in Harran but also in Ur, and would proclaim the worship of Sin as the state religion in all the lands of the Black-Headed people!

Touching the hem of the god's robe, day after day she prayed; then one night the god appeared to her in a dream and accepted her proposal. The Moon god, Adda-Guppi wrote, liked the idea: "Sin, lord of the gods of Heaven and Earth, for my good doings looked upon me with a smile; he heard my prayers; he accepted my vow. The wrath of his heart calmed. Toward Ehulhul, his temple in Harran, the divine residence in which his heart rejoiced, he became reconciled; and he had a change of heart." The god, Adda-Guppi wrote, accepted the deal:

> Sin, lord of the gods,
> looked with favor upon my words.
> Nabuna'id, my only son, issue of my womb,
> to the kingship he called—
> the kingship of Sumer and Akkad.
> All the lands from the border of Egypt,
> from the Upper Sea to the Lower Sea,
> in his hands he entrusted.

Both sides kept their bargain. "I myself saw it fulfilled," Adda-Guppi stated in the concluding segment of her inscriptions: Sin "honored his word which he spoke to me," causing Nabuna'id to ascend the Babylonian throne in 555 B.C.E.; and Nabuna'id kept his mother's vow to restore the Ehulhul temple in Harran, "perfecting its structure." He renewed the worship of Sin and Ningal (*Nikkal* in Akkadian)—"all the forgotten rites he made anew."

And then a great miracle, an occurrence unseen for generations, happened. The event is described in the two stelae of Nabuna'id, in which he is depicted holding an unusual staff and facing the celestial symbols of Nibiru, Earth, and the Moon **(Fig. 102)**:

FIGURE 102

This is the great miracle of Sin
that has by gods and goddesses
not happened in the land,
since days of old unknown;
That the people of the Earth

had neither seen nor found written
on tablets since the days of old:
That Sin, lord of gods and goddesses,
residing in the heavens,
has come down from the heavens—
in full view of Nabuna'id, king of Babylon.

Sin, the inscriptions report, did not return alone. According to the texts, he entered the restored Ehulhul temple in a ceremonial procession, accompanied by his spouse Ningal/Nikkal and his aide, the Divine Messenger Nusku.

The miraculous return of Sin "from the heavens" raises many questions, the first one being Where, "in the heavens," he had been for five or six decades. Answers to such questions can be given by combining the ancient evidence with the achievements of modern science and technology. But before we turn to that, it is important to examine all the aspects of the Departure, for it was not Sin alone who "became angry" and, leaving Earth, "went up to heaven."

The extraordinary celestial comings and goings described by Adda-Guppi and Nabuna'id took place while they were in Harran— a significant point because another eyewitness was present in that area at that very time; he was the Prophet Ezekiel; and he, too, had much to say on the subject.

Ezekiel, a priest of Yahweh in Jerusalem, was among the aristocracy and craftsmen who had been exiled, together with King Jehoiachin, after Nebuchadnezzar's first attack on Jerusalem in 598 B.C.E. They were taken forcefully to northern Mesopotamia, settling in the district of the Khabur River, just a short distance away from their ancestral home in Harran. And it was there that Ezekiel's famous vision of a celestial chariot had occurred. As a trained priest, he too recorded the place and the date: it was on the fifth day of the fourth month in the fifth year of the exile—594/593

B.C.E.—"when I was among the exiles on the banks of the river Khebar, that the heavens opened up and I saw visions of *Elohim*," Ezekiel stated at the very beginning of his prophecies; and what he saw, appearing in a whirlwind, flashing lights and surrounded by a radiance, was a divine chariot that could go up and down and sideways, and within it, "upon the likeness of a throne, the semblance of a man"; and he heard a voice addressing him as "Son of Man" and announcing his prophetic assignment.

The Prophet's opening statement is usually translated "visions of *God*." The term *Elohim*, which is plural, has been traditionally translated "God" in the singular, even when the Bible itself clearly treats it in the plural, as in "And *Elohim* said let *us* fashion the Adam in *our* image and after *our* likeness" (Genesis 1: 26). As readers of my books know, the biblical Adam tale is a rendering of the much more detailed Sumerian creation texts, where it was an Anunnaki team, led by Enki, that used genetic engineering to "fashion" the Adam. The term *Elohim*, we have shown over and over again, referred to the Anunnaki; and **what Ezekiel reported was that he had encountered an Anunnaki celestial craft**—near Harran.

The celestial craft that was seen by Ezekiel was described by him, in the opening chapter and thereafter, as the God's *Kavod* ("That which is heavy")—the very same term used in *Exodus* to describe the divine vehicle that had landed on Mount Sinai. The craft's description rendered by Ezekiel has inspired generations of scholars and artists; the resulting depictions have changed with time, as our own technology of flight vehicles has advanced. Ancient texts refer both to spacecraft and aircraft, and describe Enlil, Enki, Ninurta, Marduk, Thoth, Sin, Shamash, and Ishtar, to name the most prominent, as gods who possessed aircraft and could roam Earth's skies—or engage in aerial battles, as between Horus and Seth or Ninurta and Anzu (not to mention the Indo-European gods). Of all the varied textual descriptions and pictorial depictions of the "celestial boats" of the gods, the most appropriate to Ezekiel's vision of a Whirlwind appears to be the "whirlwind chariot"

FIGURE 103

depicted at a site in Jordan **(Fig. 103)** from which the Prophet Elijah was taken up to heaven. Helicopterlike, it had to serve just as a shuttlecraft to where full-fledged spacecraft were stationed.

Ezekiel's mission was to prophesy and warn his exiled compatriots of the coming Day of Judgment for all the nations' injustices and abominations. Then, a year later, the same "semblance of a man" appeared again, put out a hand, grabbed him, and carried him all the way to Jerusalem, to prophecy there. The city, it will be remembered, went through a starving siege, a humiliating defeat, wanton looting, a Babylonian occupation, and the exile of the king and all the nobility. Arriving there, Ezekiel saw a scene of complete breakdown of the rule of law and of religious observances. Wondering what was going on, he heard the remnant sitting in mourning, bewailing (8: 12; 9: 9):

> **Yahweh sees us no more,**
> **Yahweh has left the Earth!**

This was, we suggest, the reason why Nebuchadnezzar dared attack Jerusalem again and destroy Yahweh's temple. It was an outcry virtually identical to what Adda-Guppi had reported from Harran: "Sin, the lord of the gods, became angry with his city and his people, and went up to heaven; and the city and the people in it went to ruin."

One cannot be certain how or why events occurring in northern Mesopotamia gave rise to a notion in distant Judea that Yahweh, too, had left the Earth, but it is evident that word that God and gods departed had spread far and wide. Indeed, tablet VAT 7847, which we mentioned earlier in connection with the solar eclipse, states the following in a prophetic section regarding calamities that last 200 years:

> *Roaring the gods, flying,*
> *from the lands will go away,*
> *from the people they will be separated.*
> *The people will the gods' abodes leave in ruins.*
> *Compassion and well-being will cease.*
> *Enlil, in anger, will lift himself off.*

Like several other documents of the "Akkadian Prophecies" genre, scholars deem this text, too, a "post-event prophecy"—a text that uses events that had already happened as the basis for predicting other future events. Be that as it may, we have here a document that considerably expands the divine exodus: the angered gods, *led by Enlil*, flew away from their lands; it was not just Sin who was angered and left.

There is yet another document. It is classified by scholars as belonging to "Prophecy in Neo-Assyrian sources," though its very first words suggest authorship by a (Babylonian?) worshipper of Marduk. Here is, in full, what it says:

> *Marduk, the Enlil of the gods, got angry. His mind became furious.*
> *He made an evil plan to disperse the land and its peoples.*
> *His angry heart was bent on levelling the land and*
> *   destroying its people.*
> *A grievous curse formed in his mouth.*
> *Evil portents indicating the disruption of heavenly harmony*
> *   started appearing abundantly in heaven and on Earth.*

*The planets in the Ways of Enlil, Anu and Ea worsened
their positions and repeatedly disclosed abnormal omens.
Arahtu, the river of abundance, became a raging current.
A fierce surge of water, a violent flood like the Deluge
swept away the city, its houses and sanctuaries, turning
them to ruins.*
**The gods and goddesses became afraid, abandoned their
shrines, flew off like birds and ascended to heaven.**

What is common to all these texts are the assertions that (a) the gods grew angry with the people, (b) the gods "flew away like birds," and (c) they ascended to "heaven." We are further informed that the departure was accompanied by unusual celestial phenomena and some terrestrial disturbances. These are aspects of the Day of the Lord as prophesied by the biblical Prophets: **The Departure was related to the Return of Nibiru—the gods left Earth when Nibiru came.**

The VAT 7847 text includes an intriguing reference to a calamitous period of two centuries. The text does not make it clear whether that was a prediction of what is to follow the gods' departure, or whether it was during such a time that their anger and disappointment with Mankind grew, leading to the Departure. It seems that the latter is the case, for it is probably no coincidence that the era of biblical prophecy regarding the nations' sins and the coming judgment on the Day of the Lord began with Amos and Hosea circa 760/750 B.C.E.—two centuries before the Return of Nibiru! For two centuries the Prophets, from the only legitimate place of the "Bond Heaven-Earth"—Jerusalem—called for justice and honesty among people and peace among nations, scorned meaningless offerings and worship of lifeless idols, denounced wanton conquests and pitiless destruction, and warned one nation after another—Israel included—of the inevitable punishments, but to no avail.

If that was the case, then what had taken place was a gradual buildup of divine anger and disappointment, and the reaching of a conclusion by the Anunnaki that "enough is enough"—it was time to leave. It all brings to mind the decision of the gods, led by the disappointed Enlil, to keep the coming Deluge and the gods' lofting themselves in their celestial craft a secret from Mankind; now, as Nibiru was again nearing, it was the Enlilite gods who planned the Departure.

Who left, how did they leave, and where did they go if Sin could come back in a few decades? For the answers, let us roll the events back to the beginning.

When the Anunnaki, led by Ea/Enki, had first come to Earth to obtain the gold with which to protect their planet's endangered atmosphere, they planned to extract the gold from the waters of the Persian Gulf. When that did not work, they shifted to mining operations in southeastern Africa and smelting and refining in the E.DIN, the future Sumer. Their number increased to 600 on Earth plus 300 Igigi who operated celestial craft to a way station on Mars, from which the long-haul spacecraft to Nibiru could be launched more easily. Enlil, Enki's half-brother and rival for the succession, came and was put in overall command. When the Anunnaki toiling in the mines mutinied, Enki suggested that a "Primitive Worker" be fashioned; it was done by genetically upgrading an existing Hominid. And then the Anunnaki began to "take the daughters of the Adam as wives and had children by them" (*Genesis* 6), with Enki and Marduk breaking the taboo. When the Deluge came, the outraged Enlil said "let mankind perish," for "the wickedness of Man was great on the Earth." But Enki, through a "Noah," frustrated the plan. Mankind survived, proliferated, and in time was granted civilization.

The Deluge that swept over the Earth flooded the mines in Africa, but exposed a mother lode of gold in the Andes Mountains of South America, enabling the Anunnaki to obtain more gold more easily and quickly, and without the need for smelting and refining, for the Placer Gold—pure gold nuggests washed down from the mountains—needed only panning and collecting. It also made it possible to reduce the number of Anunnaki needed on Earth. On their state visit to Earth circa 4000 B.C.E., Anu and Antu visited the post-Diluvial gold land on the shores of Lake Titicaca.

The visit served as an opportunity to begin reducing the number of Nibiruans on Earth; it also approved peace arrangements between the rival half-brothers and their warring clans. But while Enki and Enlil accepted the territorial divisions, Enki's son Marduk never gave up the strife for supremacy that included control of the olden space-related sites. It was then that the Enlilites began to prepare alternative spaceport facilities in South America. When the post-Diluvial spaceport in the Sinai was wiped out with nuclear weapons in 2024 B.C.E., the facilities in South America were the only ones left entirely in Enlilite hands.

And so, when the frustrated and disgusted Anunnaki leadership decided that it was time to leave, some could use the Landing Place; others, perhaps with a last large haul of gold, had to use the South American facilities, near the place where Anu and Antu stayed during their visit to the area.

As earlier mentioned, the place—now called Puma-Punku—is a short distance from a shrunken Lake Titicaca (shared by Peru and Bolivia), but was then situated on the lake's southern shore, with harbor facilities. Its main remains consist of a row of four collapsed structures, each made of a single hollowed-out giant boulder **(Fig. 104)**. Each such hollowed-out set of chambers was completely inlaid inside with gold plates, held in place by gold nails—an incredible treasure hauled off by the Spaniards when they arrived in the sixteenth century. How such dwellings were so precisely hollowed out of the rocks and how four huge rocks were brought to the site remain a mystery.

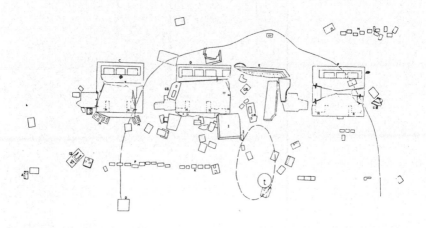

FIGURE 104

There is yet another mystery at the site. The archaeological finds in the place included a large number of unusual stone blocks that were precisely cut, grooved, angled, and shaped; some of them are shown in **Fig. 105**. One does not need an engineering degree to realize that these stones were cut, drilled, and shaped by someone with incredible technological ability and sophisticated equipment; indeed, one would doubt whether *stones* could be so shaped nowadays. The puzzle is compounded by the mystery of what purpose did these technological miracles serve; obviously, for some unknown yet highly sophisticated purpose. If it was to serve as casting dies for complex instruments, what—and whose—were those instruments?

Clearly, one can think only of the Anunnaki as possessing

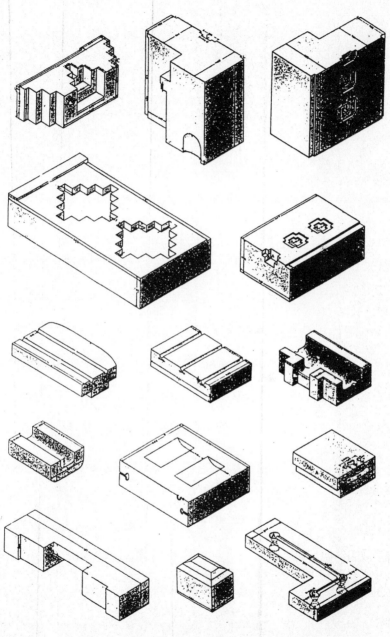

FIGURE 105

both the technology to make those "dies" and to use them or their end products. The main outpost of the Anunnaki was situated a few miles inland, at a site now known as Tiwanacu (earlier spelled Tiahuanacu), now belonging to Bolivia. One of the first European explorers to reach it in modern times, George Squier, described the place in his book *Peru Illustrated* as "The Baalbec of the New world"—a comparison more valid than he realized.

The next main modern explorer of Tiwanaku, Arthur Posnansky (*Tihuanacu—The Cradle of American Man*), reached astounding conclusions regarding the site's age. The principal aboveground structures in Tiwanaku (there are numerous subterranean ones) include the **Akapana**, an artificial hill riddled with channels, conduits, and sluices whose purpose is discussed in *The Lost*

FIGURE 106

Realms. A tourist favorite is a stone gateway known as the **Gate of the Sun**, a prominent structure that was also cut from a single boulder, with some of the precision exhibited at Puma-Punku. It probably served an astronomical purpose and undoubtedly a calendrical one, as the carved images on the archway indicate; those carvings are dominated by the larger image of the god Viracocha holding the lightning weapon that clearly emulated the Near Eastern Adad/Teshub **(Fig. 106)**. Indeed, in *The Lost Realms* I have suggested that he *was* Adad/Teshub.

The Gate of the Sun is so positioned that it forms an astronomical observation unit with the third prominent structure at Tiwanaku, called the **Kalasasaya**. It is a large rectangular structure with a sunken central courtyard and is surrounded by standing stone pillars. Posnansky's suggstion that the Kalasasaya served as an observatory has been confirmed by subsequent explorers; his conclusion, based on Sir Norman Lockyer's archaeoastronomy guidelines, that the astronomical alignments of the Kalasasaya show that it was built thousands of years before the Incas was so incredible that German astronomical institutions sent teams of scientists to check this out. Their report, and subsequent additional verifications (viz. the scientific journal *Baesseler Archiv*, volume 14) affirmed that the Kalasasaya's orientation unquestionably matched the Earth's obliquity either in 10,000 B.C.E. **or 4000 B.C.E.**

Either date, I wrote in *The Lost Realms*, was fine with me— the earlier soon after the Deluge, when the gold-obtaining operations began there, or the later date, when Anu visited; both dates matched the activities of the Anunnaki there, and the evidence for the presence of the Enlilite gods is all over the place.

Archaeological, geological, and mineralogical research at the site and in the area confirmed that Tiwanaku also served as a metallurgical center. Based on various finds and the images on the Gate of the Sun **(Fig. 107a)** and their similarity to depictions in ancient Hittite sites in Turkey **(Fig. 107b)**, I have suggested

FIGURE 107a

FIGURE 107b

that the gold (and tin!) obtainment operations there were super-vised by Ishkur/Adad, Enlil's youngest son. His domain in the Old World was Anatolia, where he was worshipped by the Hit-tites as Teshub, the "weather god" whose symbol was the light-ning rod; such a huge symbol, enigmatically carved on a steep mountainside **(Fig. 108)**, can be seen from the air or from out in the ocean in the Bay of Paracas, Peru, a natural harbor downhill from Tiwanaku. Nicknamed the Candelabra, the symbol is 420 feet long and 240 feet wide, and its lines, which are 5 to 15 feet wide, have been etched into the hard rocks to a depth of about 2 feet—and no one knows by whom and when or how, unless it was Adad himself who wanted to declare his presence.

To the north of the bay, inland in the desert between the In-genio and Nazca rivers, explorers have found one of the most puz-

FIGURE 108

zling riddles of antiquity, the so-called **Nazca Lines**. Called by some "the world's largest artworks," a vast area (some 200 square miles!) that extends eastward from the *pampa* (flat desert) to the rugged mountains was used by "someone" as a canvas to draw on it scores of images; the drawings are so huge that they make no sense at ground level—but when viewed from the air, clearly represent known and imaginary animals and birds **(Fig. 109)**. The drawings were made by removing the topsoil to a depth of several inches, and were executed with a unicursal line—a continuous line that curves and twists without crossing over itself. Anyone flying over the area (there are small planes at the service of tourists there) invariably concludes that "someone" *airborne* has used a soil-blasting device to doodle on the ground below.

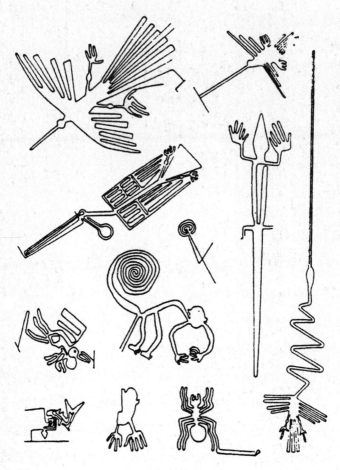

FIGURE 109

Directly relevant to the subject of the Departure, however, is another even more puzzling feature of the Nazca Lines—actual *"lines" that look like wide runways* (**Fig. 110**). Straight without fault, these flat stretches—sometimes narrow, sometimes wide, sometimes short, sometimes long—run straight over hills and vales, no matter the shape of the terrain. There are some 740 straight "lines," sometimes combined with triangular "trapezoids" (**Fig. 111**). They frequently criss-cross each other without rhyme or reason, sometimes running over the animal drawings, revealing that the lines were made at different times.

FIGURE 110

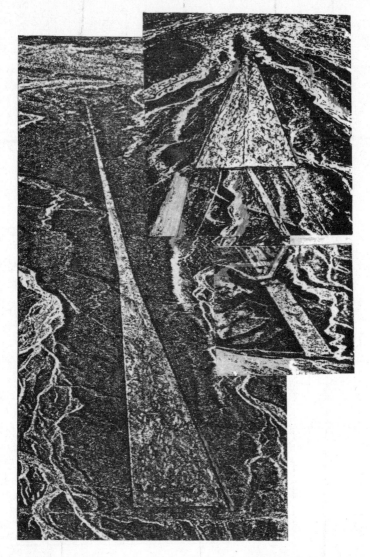

FIGURE III

Various attempts to resolve the mystery of the Lines, including those by the late Maria Reiche, who made it her lifelong project, failed whenever an explanation was sought in terms of "it was done by native Peruvians"—people of a "Nazca culture" or a "Paracas civilization" or the likes. Studies (including some by the National Geographic Society) aimed at uncovering astronomical orientations for the lines—alignments with solstices, equinoxes, this or that star—led nowhere. For those who rule out an "Ancient Astronauts" solution, the enigma remains unresolved.

Though the wider lines look like airport runways, on which wheeled aircraft roll to take off (or to land), this is not the case here, if only because the "lines" are not horizontally level—they run straight over uneven terrain, ignoring hills, ravines, and gullies. Indeed, rather than being there to enable takeoff, they appear to be the *result of takeoffs* by craft taking off and leaving on the ground below "lines" created by their engine's exhaust. That the "celestial chambers" of the Anunnaki did emit such exhausts is indicated by the Sumerian pictograph (read DIN.GIR) for the space gods **(Fig. 112)**.

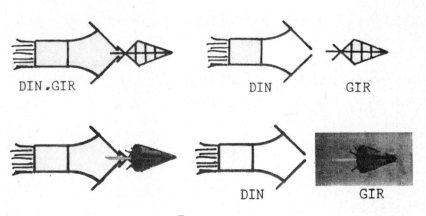

FIGURE 112

This, I suggest, is the solution of the puzzle of the "Nazca Lines": Nazca was the last spaceport of the Anunnaki. It served them after the one in the Sinai was detroyed, and then it served them for the final Departure.

There are no eyewitness-report texts regarding the airborne craft and flights in Nazca; there are, as we have shown, texts from Harran and Babylon regarding the flights that undoubtedly used the Landing Place in Lebanon. The eyewitness reports relating to those departure flights and Anunnaki's craft include the testimony of the Prophet Ezekiel and the inscriptions of Adda-Guppi and Nabunaid.

The inevitable conclusion must be that from at least 610 B.C.E. through probably 560 B.C.E., the Anunnaki gods were methodically leaving planet Earth.

Where did they go as they lifted off Earth? It had to be, of course, a place from which Sin could return relatively soon once he changed his mind. The place was the good old Way Station on Mars, from which the long-distance spaceships raced to intercept and land on the orbiting Nibiru.

As detailed in *The Twelfth Planet*, Sumerian knowledge of our Solar system included references to the use of Mars by the Anunnaki as a Way Station. It is evidenced by a remarkable depiction on a 4,500-year-old cylinder seal now in the Hermitage Museum in St. Petersburg, Russia **(Fig. 113)** that shows an astronaut on Mars (the sixth planet) communicating with one on Earth (the seventh planet, counting from the outside in), with a spacecraft in the heavens between them. Benefiting from Mars's lower gravity compared to that of Earth, the Anunnaki had found it easier and more logical to first transport themselves and their car-

FIGURE 113

gos in shuttlecraft from Earth to Mars, and there transfer to reach Nibiru (and vice versa).

In 1976, when all that was first presented in *The Twelfth Planet*, Mars was still held to be an airless, waterless, lifeless, hostile planet, and the suggestion that a space base had once existed there was deemed by establishment scholars even more far out than the notion of "Ancient Astronauts." By the time *Genesis Revisited* was published in 1990, there were enough of NASA's own findings and photographs from Mars to fill up a whole chapter titled "A Space Base on Mars." The evidence showed that Mars once had water, and included photographs of walled structures, roads, a hublike compound **(Fig. 114** shows just two such photographs)—and the famous Face **(Fig. 115)**.

Both the United States and the Soviet Union (now Russia) made great efforts to reach and explore Mars with unmanned spacecraft; unlike other space endeavors, the missions to Mars— since augmented by the European Union—have met with an unusual, troubling, and puzzling high rate of failures, including bewildering unexplained disappearances of spacecraft. But due to persistent efforts, enough U.S., Soviet, and European unmanned spacecraft have managed to reach and explore Mars in the last two decades, that by now the scientific journals—of the same "Doubting Thomases" of the 1970s—have been filled with reports, studies, and photographs announcing that Mars did have a sizeable and still has a thin atmosphere; that it once had rivers, lakes, and oceans and still has water, at some places just below the surface and in some instances even visible as small frozen lakes—as a medley of the headlines shows **(Fig. 116)**. In 2005 NASA's Mars Rovers sent back chemical and photographic evidence backing those conclusions; together with some of the Rovers' amazing photographs showing structural remains—like a sand-covered wall with distinct right-angled corners **(Fig. 117)**—they should suffice here to make the point: **Mars could, and did, serve as a Way Station for the Anunnaki.**

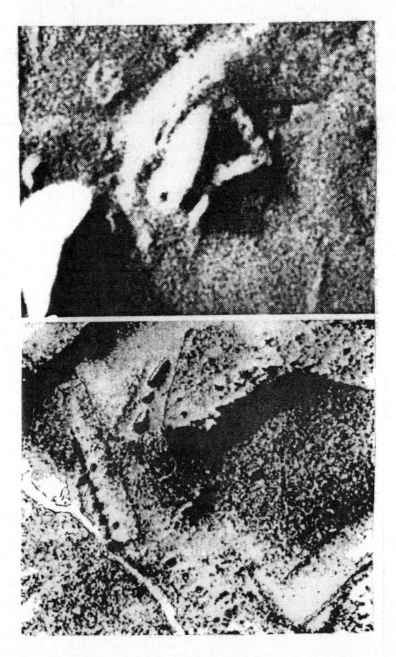

FIGURE 114

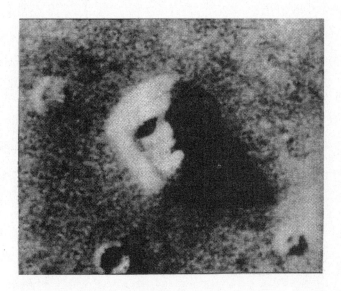

FIGURE 115

FIGURE 116

FIGURE 117

It was the first close-by destination of the departing gods, as confirmed by the relatively quick return of Sin. Who else left, who stayed behind, who might return?

Surprisingly, some of the answers also come from Mars.

Chapter XIV

THE END OF DAYS

Mankind's recollection of landmark events in its past—"legends" or "myths" to most historians—includes tales deemed "universal" in that they have been part of the cultural or religious heritage of peoples all over the Earth. Tales of a First Human Couple, of a Deluge, or of gods who came from the heavens, belong to that category. So do tales of the gods' departure back to the heavens.

Of particular interest to us are such collective memories by the peoples and in the lands where the departures had actually taken place. We have already covered the evidence from the ancient Near East; it also comes from the Americas, and it embraces both Enlilite and Enki'ite gods.

In South America, the dominant deity was called *Viracocha* ("Creator of All"). The Aymara natives of the Andes told of him that his abode was in Tiwanaku, and that he gave the first two brother-sister couples a golden wand with which to find the right place to establish Cuzco (the eventual Inca capital), the site for the observatory of Machu Picchu, and other sacred sites. And then, having done all that, *he left*. The grand layout, which simulated a square ziggurat with its corners oriented to the cardinal points, then marked the direction of his eventual departure **(Fig. 118)**. We have identified the god of Tiwanaku as Teshub/Adad of the Hittite/Sumerian pantheon, Enlil's youngest son.

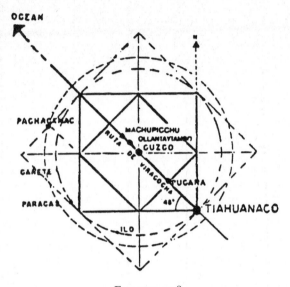

FIGURE 118

In Mesoamerica, the giver of civilization was the "Winged Serpent" *Quetzalcoatl*. We have identified him as Enki's son Thoth of the Egyptian pantheon (Ningishzidda to the Sumerians) and as the one who, in 3113 B.C.E., brought over his African followers to establish civilization in Mesoamerica. Though the time of his departure was not specified, it had to coincide with the demise of his African protégés, the Olmecs, and the simultaneous rise of the native Mayas—circa 600/500 B.C.E. The dominant legend in Mesoamerica was his promise, when he departed, *to return*—on the annivesray of his Secret Number 52.

And so it was, by the middle of the first millennium B.C.E., in one part of the world after another, that Mankind found itself without its long-worshipped gods; and before long, the question (which has been asked by my readers) began to preoccupy Mankind: *Will they return?*

Like a family suddenly abandoned by its father, Mankind grasped for the hope of a Return; then, like an orphan needing help, Mankind cast about for a Savior. The Prophets promised it will surely happen—at the *End of Days*.

At the peak of their presence, the Anunnaki numbered 600 on Earth plus another 300 Igigi based on Mars. Their number was falling after the Deluge and especially after Anu's visit circa 4000 B.C.E. Of the gods named in the early Sumerian texts and in long God Lists, few remained as the millennia followed each other. Most returned to their home planet; some—in spite of their wonted "immortality"—died on Earth. We can mention the defeated Zu and Seth, the dismembered Osiris, the drowned Dumuzi, the nuclear-afflicted Bau. The departures of the Anunnaki gods as Nibiru's return loomed were the dramatic finale.

The awesome times when the gods resided in sacred precincts in the people's cities, when a Pharaoh claimed that a god was riding along in his chariot, when an Assyrian king boasted of help from the skies, were over and gone. Already in the days of the Prophet Jeremiah (626–586 B.C.E.), the nations surrounding Judea were mocked for worshipping not a "living god" but idols made by craftsmen of stone, wood, and metal—gods who needed to be carried, for they could not walk.

With the final departure taking place, who of the great Anunnaki gods remained on Earth? To judge by who was mentioned in the texts and inscriptions from the ensuing period, we can be certain only of Marduk and Nabu of the Enki'ites; and of the Enlilites, Nannar/Sin, his spouse Ningal/Nikkal and his aide Nusku, and probably also Ishtar. On each side of the great religious divide there was now just one sole Great God of Heaven and Earth: Marduk for the Enki'ites, Nannar/Sin for the Enlilites.

The story of Babylonia's last king reflected the new circumstances. He was chosen by **Sin** in his cult-center Harran—but he required the consent and blessing of **Marduk** in Babylon, and the celestial confirmation by the appearance of Marduk's planet; and he bore the name **Nabu**-Na'id. This divine co-regnum might have been an attempt at Dual Monotheism (to coin an expression); but *its **unintended consequence was to plant the seeds of** Islam*.

The historical record indicates that neither gods nor people were happy with these arrangements. Sin, whose temple in Harran was restored, demanded that his great ziggurat temple in Ur should also be rebuilt and become the center of worship; and in Babylon, the priests of Marduk were up in arms.

A tablet now in the British Museum is inscribed with a text that scholars have titled *Nabunaid and the Clergy of Babylon*. It contains a list of accusations by the Babylonian priests against Nabunaid. The charges ran from civil matters ("law and order are not promulgated by him"), through neglect of the economy ("the farmers are corrupted," "the traders' roads are blocked"), and lack of public safety ("nobles are killed"), to the most serious charges: religious sacrilege—

> He made an image of a god which nobody had seen before in
>   the land.
> He placed it in the temple, raised it upon a pedestal,
> He called it by the name of Nannar, with lapis lazuli he
>   adorned it,
> Crowned it with a tiara in the shape of an eclipsed moon,
> Made for its hand the gesture of a demon.

It was, the accusations continued, a strange statue of a deity, never seen before, "with hair reaching down to the pedestal." It was so unusual and unseemly, the priests wrote, that even Enki and Ninmah (who ended up with strange chimera creatures when they attempted to fashion Man) "could not have conceived it"; it was so strange that "not even the learned Adapa"—an icon of utmost human knowledge—"could have named it." To make matters worse, two unusual beasts were sculpted as its guardians— one a "Deluge demon" and the other a wild bull; then the king took this abomination and placed it in Marduk's Esagil temple. Even more offending was Nabunaid's announcment that hence-forth the *Akitu* festival, during which the near-death, resurrec-tion, exile, and final triumph of Marduk were reenacted, would no longer be celebrated.

Declaring that Nabunaid's "protective god became hostile to him" and that "the former favorite of the gods was now fated to misfortune," the Babylonian priests forced Nabunaid to leave Bab-ylon and go into exile "in a distant region." It is a historical fact that Nabunaid indeed left Babylon and named his son Bel-Shar-Uzur—the Belshazzar of the biblical Book of Daniel—as regent.

The "distant region" to which Nabunaid went in self-exile was Arabia. As various inscriptions attest, his entourage included Jews from among the Judean exiles in the Harran region. His principal base was at a place called Teima, a caravan center in what is now northwestern Saudi Arabia that is mentioned several times in the Bible. (Recent excavations there have uncovered cuneiform tablets attesting to Nabunaid's stay.) He established six other settlements for his followers; five of the towns were listed— a thousand years later—by Arabian writers as Jewish towns. ***One of them was Medina, the town where Muhammed founded Islam.***

The "Jewish angle"in the Nabunaid tale has been reinforced by the fact that a fragment of the Dead Sea scrolls, found at Qumran on the shores of the Dead Sea, mentions Nabunaid and asserts that he was suffering in Teima from an "unpleasant skin disease" that was cured only after "a Jew told him to give honor to the God Most High." All that has led to speculation that Nabunaid was contem-plating Monotheism; but to him the God Most High was not the Judeans' Yahweh, but his benefector Nannar/Sin, the Moon god, whose crescent symbol has been adopted by Islam; and there is little doubt that its roots can be traced back to Nabunaid's stay in Arabia.

Sin's whereabouts fade out of Mesopotamian records after the time of Nabunaid. Texts discovered at Ugarit, a "Canaanite" site on the Mediterranean coast in Syria now called Ras Shamra, describe the Moon god as retired, with his spouse, to an oasis at the confluence of two bodies of water, "near the cleft of the two seas." Ever wondering why the Sinai peninsula was named in honor of Sin and its main central crossroads in honor of his spouse Nikkal (the place is still called, in Arabic, Nakhl), I surmised that the aged couple retired to somewhere on the shores of the Red Sea and the Gulf of Eilat.

The Ugaritic texts called the Moon god *EL*—simply, "God," a forerunner of Islam's *Allah*; and his moon-crescent symbol crowns every Moslem mosque. And as tradition demands, the mosques are flanked, to this day, by minarets that simulate multistage rocketships ready to be launched (Fig. 119).

FIGURE 119

The last chapter in the Nabunaid saga was linked to the emergence on the scene of the ancient world of the **Persians**—a name given to a medley of peoples and states on the Iranian plateau that included the olden Sumerian Anshan and Elam and the land of the later Medes (who had a hand in the demise of Assyria).

It was in the sixth century B.C.E. that a tribe called **Achaemeans** by Greek historians who recorded their deeds emerged from the northern outskirts of those territories, seized control, and unified them all to become a mighty new empire. Though deemed to racially be "Indo-Europeans," their tribal name stemmed from that of their ancestor *Hakham-Anish,* which meant "Wise Man" in Semitic Hebrew—a fact that some attribute to the influence of Jewish exiles from the Ten Tribes who had been relocated to that region by the Assyrians. Religiously, the Achaemean Persians apparently adopted a Sumerian-Akkadian pantheon akin to its Hurrian-Mitannian version, which was a step to the Indo-Aryan one of the Sanskrit Vedas—a mixture that is conveniently simplified by just stating that they believed in a God Most High whom they called *Ahura-Mazda* ("Truth and Light").

In **560 B.C.E.** the Achaemean king died and his son Kurash succeeded him on the throne and made his mark on subsequent historic events. We call him *Cyrus*; the Bible called him Koresh and considered him Yahweh's emissary for conquering Babylon, overthrowing its king, and rebuilding the destroyed Temple in Jerusalem. "Though you knowest Me not, I, Yahweh, the God of Israel, am thy caller who hath called you by name . . . who will help you though you don't recognize me," the biblical God stated through the prophet Isaiah (44: 28 to 45: 1–4).

That end of Babylonian kingship was most dramatically foretold in the *Book of Daniel*. One of the Judean exiles taken to Babylon, Daniel was serving in the Babylonian court of Belshazzar when, during a royal banquet, a floating hand appeared and wrote on the wall *MENE MENE TEKEL UPHARSIN*. Astounded and mystified, the king called his wizards and seers to decipher the inscription, but none could. As a last resort, the exiled Daniel was called in, and he told the king the inscription's meaning: God has weighed Babylon and its king and, finding them wanting, numbered their days; they will meet their end by the hand of the Persians.

In 539 B.C.E. Cyrus crossed the Tigris River into Babylonian territory, advanced on Sippar where he intercepted a rushing-back Nabunaid, and then—claiming that Marduk himself had invited him—entered Babylon without a fight. Welcomed by the priests who considered him a savior from the heretic Nabunaid and his disliked son, Cyrus "grasped the hands of Marduk" as a sign of homage to the god. But he also, in one of his very first proclamations, rescinded the exile of the Judeans, permitted the rebuilding of the Temple in Jerusalem, and ordered the return of all the Temple's ritual objects that were looted by Nebuchadnezzar.

The returning exiles, under the leadership of Ezra and Nehemiah, completed the rebuilding of the Temple—henceforth known as the Second Temple—in 516 B.C.E.—exactly, as was prophesied by Jeremiah, seventy years after the First Temple was destroyed. The Bible considered Cyrus an instrument of God's plans, an "anointed of Yahweh"; historians believe that Cyrus proclaimed a general religious amnesty that allowed each people to worship as they pleased. As to what Cyrus himself might have believed, to judge by the monument he had erected for himself, he appears to have envisioned himself as a winged Cherub **(Fig. 120)**.

Cyrus—some historians attach the epithet "the great" to his name—consolidated into a vast Persian empire all the lands that had once been Sumer & Akkad, Mari and Mittani, Hatti and Elam, Babylonia and Assyria; it was left to his son Cambyses (530–522 B.C.E.) to extend the empire to Egypt. Egypt was just recovering from a period of disarray that some consider a Third Intermediate Period, during which it was disunited, changed capitals several times, was ruled by invaders from Nubia, or had no

FIGURE 120

central authority at all. Egypt was also in disarray religiously, its priests uncertain who to worship, so much so that the leading cult was that of the dead Osiris, the leading deity the female Neith whose title was *Mother of God*, and the principal "cult object" a bull, the sacred Apis Bull, for whom elaborate funerals were held. Cambyses, too, like his father, was no religious zealot, and let people worship as they pleased; he even (according to an inscribed stela now in the Vatican museum) learnt the secrets of the worship of Neith and participated in a ceremonial funeral of an Apis bull.

These religious *laissez-faire* policies bought the Persians peace in their empire, but not forever. Unrest, uprisings, and rebellions kept breaking out almost everywhere. Especially troublesome were growing commercial, cultural, and religious ties between Egypt and Greece. (Much information about that comes from the Greek historian Herodotus, who wrote extensively about Egypt after his visit there circa 460 B.C.E., coinciding with the beginning of Greece's "golden age.") The Persians could not be pleased with those ties, above all because Greek mercenaries were participating in the local uprisings. Of particular concern were also the provinces in Asia Minor (present-day Turkey), at the western tip of which Asia and the Persians faced Europe and the Greeks. There, Greek settlers were reviving and reinforcing olden settlements; the Persians, on their part, sought to ward off the troublesome Europeans by seizing nearby Greek islands.

The growing tensions broke into open warfare when the Persians invaded the Greek mainland and were beaten at Marathon in 490 B.C.E.. A Persian invasion by sea was beaten off by the Greeks in the straits of Salamis a decade later, but the skirmishes and battles for control of Asia Minor continued for another century, even as in Persia king followed king and in Greece Athenians, Spartans, and Macedonians fought one another for supremacy.

In those double struggles—one among the mainland Greeks, the other with the Persians—the support of the Greek settlers of Asia Minor was very important. No sooner did the Macedonians win the upper hand on the mainland than their king, Philip II, sent an armed corps over the Straits of Hellespont (today's Dardanelles) to secure the loyalty of the Greek settlements. In 334 B.C.E. his successor, Alexander ("the Great"), heading an army 15,000 strong, crossed into Asia at the same place and launched a major war against the Persians.

Alexander's astounding victories and the resulting subjugation of the Ancient East to Western (Greek) domination have been told and retold by historians—starting with some who had accompanied Alexander—and need no repetition here. What does need to be described are the *personal* reasons for Alexander's foray into Asia and Africa. For, apart from all geopolitical or economic reasons for the Greek-Persian great war, there was Alexander's own personal quest: there had been persistent rumors in the Macedonian court that not King Philip but a god—an Egyptian god—was Alexander's true father, having come to the queen, Olympias, disguised as a man. With a Greek pantheon that derived from across the Mediterranean Sea and headed (like the Sumerian twelve) by twelve Olympians, and with tales of the gods ("myths") that emulated the Near Eastern tales of the gods, the appearance of one such god in the Macedonian court was not deemed an impossibility. With court shenanigans that involved a young Egyptian mistress of the king and marital strife that included divorce and murders, the "rumors" were believed—first and foremost by Alexander himself.

A visit by Alexander to the oracle in Delphi to find out whether he was indeed the son of a god and therefore immortal only intensified the mystery; he was advised to seek an answer at an Egyptian sacred site. It was thus that as soon as the Persians were beaten in the first battle, Alexander, rather than pursuing them, left his main army and rushed to the oasis of Siwa in Egypt. There the priests assured him that he indeed was a demigod, the son of the ram-god Amon. In celebration, Alexander issued silver coins showing him with ram's horns **(Fig. 121)**.

But what about the immortality? While the course of the resumed warfare and Alexander's conquests have been documented by his campaign historian Callisthenes and other historians, his personal quest for Immortality is mostly known from sources deemed to be pseudo-Callisthenes, or "Alexander Romances" that embellished fact with legend. As detailed in *The Stairway to Heaven*, the Egyptian priests directed Alexander from Siwa to Thebes. There, on the Nile River's western shore, he could see

FIGURE 121

in the funerary temple built by Hatshepsut the inscription attesting to her being fathered by the god Amon when he came to her mother disguised as the royal husband—exactly like the tale of Alexander's demigod conception. In the great temple of Ra-Amon in Thebes, in the Holy of Holies, Alexander was crowned as a Pharaoh. Then, following the directions given in Siwa, he entered subterranean tunnels in the Sinai peninsula, and finally he went to where Amon-Ra, alias Marduk, was—to Babylon. Resuming the battles with the Persians, Alexander reached Babylon in 331 B.C.E., and entered the city riding in his chariot.

In the sacred precinct he rushed to the Esagil ziggurat temple to grasp the hands of Marduk as conquerors before him had done. *But the great god was dead.*

According to the pseudo-sources, Alexander saw the god lying in a golden coffin, his body immersed (or preserved) in special oils. True or not, the facts are that *Marduk was no longer alive,* and that his Esagil ziggurat was, without exception, described as his **tomb** by subsequent established historians.

According to Diodorus of Sicily (first century B.C.E.), whose *Bibliothca historica* is known to have been compiled from verified reliable sources, "scholars called Chaldaeans, who have gained a great reputation in astrology and who are accustomed to predict future events by a method based on age-old observations," warned Alexander that he would die in Babylon, but "could escape the danger if he re-erected *the tomb of Belus* which had been demolished by the Persians" (Book XVII, 112.1). Entering the city anyway, Alexander had neither the time nor the manpower to do the repairs, and indeed died in Babylon in 323 B.C.E.

The first century **B.C.E.** historian-geographer Strabo, who was born in a Greek town in Asia Minor, described Babylon in his famed *Geography*—its great size, the "hanging garden" that was one of the Seven Wonders of the World, its high buildings constructed of baked bricks, and so on, and said this in section 16.I.5 (emphasis added):

Here too is the tomb of Belus, now in ruins,
having been demolished by Xerxes, as it is said.
It was a quadrangular pyramid of baked bricks,
not only being a stadium in height,
but also having sides a stadium in length.
Alexander intended to repair this pyramid;
but it would have been a large task
and would have required a long time,
so that he could not finish what he had attempted.

According to this source, the **tomb** of *Bel*/Marduk was destroyed by Xerxes, who was the Persian king (and ruler of Babylon) from 486 to 465 B.C.E. Strabo, in Book 5, had earlier stated that *Belus* was lying in a coffin when Xerxes decided to destroy the temple, in 482 B.C.E. Accordingly, Marduk died not long before (Germany's leading Assyriologists, meeting at the University of Jena in 1922, concluded that Marduk was already in his tomb in 484 B.C.E.). Marduk's son Nabu also vanished from the pages of history about the same time. *And thus came to an end, an almost human end, the saga of the gods who shaped history on planet Earth.*

That the end came as the Age of the Ram was waning was probably no coincidence, either.

With the death of Marduk and the fading away of Nabu, all the great Anunnaki gods who had once dominated Earth were gone; with the death of Alexander, the real or pretended demigods who linked Mankind to the gods were also gone. For the first time since Adam was fashioned, Man was without his creators.

In those despondent times for Mankind, hope came forth from Jerusalem.

Amazingly, the story of Marduk and his ultimate fate in

Babylon had been correctly foretold in biblical prophecies. We have already noted that Jeremiah, while forecasting a crushing end for Babylon, made the distinction that its god Bel/Marduk was only doomed to "wither"—to remain, but to grow old and confused, to shrivel and die. We should not be surprised that it was a prophecy that came true.

But as Jeremiah correctly predicted the final downfall of Assyria, Egypt, and Babylon, he accompanied those predictions with prophecies of a reestablished Zion, of a rebuilt temple, and of a "happy end" for all nations *at the End of Days*. It would be, he said, a future that God had planned "in his heart" all along, a secret that shall be revealed to Mankind (23: 20) at a predetermined future time: *"at the End of Days you shall perceive it"* (30: 24), and "at that time, they shall call Jerusalem Yahweh's Throne, and all nations shall assemble there" (3: 17).

Isaiah, in his second set of prophecies (sometimes called the Second Isaiah), identifying Babylon's god as the "Hiding god"—which is what "Amon" meant—foresaw the future in those words:

> **Bel** *is bowed down,* **Nebo** *is cowered,*
> *their images are loads for beasts and cattle . . .*
> *Together they stoopeth, they bowed down,*
> *unable to save themselves from capture.*
>
> ISAIAH 46:1–2

These prophecies, as did Jeremiah's, also contained the promise that Mankind will be offered a new beginning, new hope; that a Messianic Time will come when "the wolf shall dwell with the lamb." And, the Prophet said, "it shall come to pass *at the End of Days* that the Mount of Yahweh's Temple shall be established as foremost of all mountains, exalted above all hills; and all the nations shall throng unto it"; it will be then that the nations "shall beat their swords into ploughshares and their spears into pruning hooks, nation shall not lift up sword against nation, and they shall no longer teach war" (Isaiah 2: 1–4).

The assertion that after troubles and tribulations, after people and nations shall be judged for their sins and transgressions, a time of peace and justice shall come was also made by the early Prophets even as they predicted the Day of the Lord as judgment day. Among them were Hosea, who foresaw the *return of the kingdom of God through the House of David at the **End of Days***, and Micha, who—using words identical to those of Isaiah—declared that "at the End of Days it shall come to pass." Significantly, Micha too considered the **restoration of God's Temple in Jerusalem** *and Yahweh's universal reign* **through a descendant of David** as a prerequisite, a "must" destined from the very beginning, "emanating from ancient times, from everlasting days."

There was thus a combination of two basic elements in those End of Days predictions: one, that the Day of the Lord, a day of judgment upon Earth and the nations, will be followed by Restoration, Renewal, and a benevolent era centered on Jerusalem. The other is, that it has all been preordained, that the End was already planned by God at the Beginning. Indeed, the concept of an End of Epoch, a time when the course of events shall come to a halt—a precursor, one may say, of the current idea of the "End of History"—and a new epoch (one is almost tempted to say, a *New Age*), a new (and predicted!) cycle shall begin, can already be found in the earliest biblical chapters.

The Hebrew term *Acharit Hayamim* (sometimes translated "last days," "latter days," but more accurately "end of days") was already used in the Bible in *Genesis* (Chapter 49), when the dying Jacob summoned his sons and said: "Gather yourselves together, that I may tell you that which shall befall you at the **End of Days**." It is a statement (followed by detailed predictions that many associate with the twelve houses of the zodiac) that presupposes prophecy by being based on advance knowledge of the future. And again, in *Deuteronomy* (Chapter 4), when Moses, before dying, reviewing

Israel's divine legacy and its future, counseled the people thus: "When you in tribulations shall be and such things shall befall you, in the **End of Days** to Yahweh thy God return and hearken to His voice."

The repeated stress on the role of Jerusalem, on the essentiality of its Temple Mount as the beacon to which all nations shall come streaming, had more than a theological-moral reason. A very practical reason is cited: the need to have the site ready for the return of Yahweh's *Kavod*—the very term used in *Exodus* and then by Ezekiel to describe God's celestial vehicle! The *Kavod* that will be enshrined in the rebuilt Temple, "from which I shall grant peace, shall be greater than the one in the First Temple," the Prophet Haggai was told. Significantly, the *Kavod*'s coming to Jerusalem was repeatedly linked in Isaiah to the other space-related site—in Lebanon: *It is from there that God's **Kavod** shall arrive in Jerusalem,* verses 35: 2 and 60: 13 stated.

One cannot avoid the conclusion that a divine Return was expected at the End of Days; but when was the End of Days due?

The question—one to which we shall offer our own answer—is not new, for it has already been asked in antiquity, even by the very Prophets who had spoken of the End of Days.

Isaiah's prophecy about the time "when a great trumpet shall be blown" and the nations shall gather and "bow down to Yahweh on the Holy Mount in Jerusalem" was accompanied by his admission that without details and timing the people could not understand the prophecy. "Precept is upon precept, precept is within precept, line is upon line, line is with line, a little here, somewhat there" was how Isaiah (28: 10) complained to God. Whatever answer he was given, he was ordered to seal and hide the document; no less than three times, Isaiah changed the word for "letters" of a script—*Otioth*—to *Ototh*, which meant "oracular signs," hinting at the existence of a kind of secret **"Bible Code"** due to which the divine plan could not be comprehended until the right time. Its secret code might have been hinted at when the Prophet asked God—identified as "Creator of the letters"—to "tell us the letters backward" (41: 23).

The prophet Zephaniah—whose very name meant "By Yahweh encoded"—relayed a message from God that it will be at the time of the nations' gathering that He "will speak in a clear language." But that said no more than saying, "You'll know when it will be time to tell."

No wonder, then, that in its final prophetic book, the Bible dealt almost exclusively with the question of WHEN—when will the End of Days come? It is the *Book of Daniel,* the very Daniel who deciphered (correctly) for Belshazzar the Writing on the Wall. It was after that that Daniel himself began to have omendreams and to see apocalyptic visions of the future in which the "Ancient of Days" and his archangels played key roles. Perplexed, Daniel asked the angels for explanations; the answers consisted of predictions of future events, taking place at, or leading to, the End of Time. And when will that be? Daniel asked; the answers, which on the face of it seemed precise, only piled up enigmas upon puzzles.

In one instance an angel answered that a phase in future events, a time when "an unholy king shall try to change the times and the laws," will last *"a time, times and a half time";* only after that will the promised Messianic Time, when "the kingdom of heaven will be given to the people by the Holy Ones of the Most High," come about. Another time the responding angel said: "Seventy sevens and seventy sixties of years have been decreed for your people and your city until the measure of transgression is filled and prophetic vision is ratified"; and yet another time that "after the seventies and sixties and two of years, the Messiah will be cut off, a leader will come who will destroy the city, and the end will come through a flood."

Seeking a clearer answer, Daniel then asked a divine messenger to speak plainly: "How long until the end of these awful things?" In response, he again received the enigmatic answer that the End will come after "*a time, times and a half time*." But what did "time, times and a half time" mean, what did "seventy weeks of years" mean?

"I heard and did not understand," Daniel stated in his book. "So I said: My lord, what will be the outcome of these things?" Again speaking in codes, the angel answered: "from the time the regular offering is abolished and an appalling abomination is set up, it will be a thousand and two hundred and ninety days; happy is the one who waits and reaches one thousand three hundred and thirty five." And having given Daniel that information, the angel—who had called him before "Son of Man"—told him: "Now, go on to thy end, and arise for your destiny at the End of Days."

Like Daniel, generations of biblical scholars, savants and theologians, astrologers and even astronomers—the famed Sir Isaac Newton among the latter—also said "we heard, but did not understand." The enigma is not just the meaning of "time, time and a half" and so on, but from when does (or did) the count begin? The uncertainty stems from the fact that the symbolic visions seen by Daniel (such as a goat attacking a ram, or two horns multiplying to four and then dividing) were explained to him by the angels as events that were to take place well beyond Babylon of Daniel's time, beyond its predicted fall, even beyond the prophesied rebuilding of the Temple after seventy years. The rise and demise of the Persian empire, the coming of the Greeks under Alexander's leadership, even the division of his conquered empire among his successors—all are foretold with such accuracy that many scholars believe that the Daniel prophecies are of the "post-event" genre—that the book's prophetic part was actually written circa 250 B.C.E. but pretended to have been written three centuries earlier.

The clinching argument is the reference, in one of the angelic encounters, to the start of the count "from the time that regular

THE END OF DAYS

offering [in the temple] is abolished and an appalling abomination is set up." That could only refer to the events that took place in Jerusalem on the 25th day of the Hebrew month Kislev in **167 B.C.E.**

The date is precisely recorded, for it was then that "the abomination of desolation" was placed in the Temple, marking—many then believed—the start of the End of Days.

Chapter XV

JERUSALEM: A CHALICE, VANISHED

In the twenty-first century B.C.E., when nuclear weapons were first used on Earth, Abraham was blessed with wine and bread at *Ur-Shalem* in the name of the God Most High—and proclaimed Mankind's first Monotheistic religion.

Twenty-one centuries later, a devout descendant of Abraham, celebrating a special supper in *Jerusalem*, carried on his back a cross—the symbol of a certain planet—to a place of execution, and gave rise to another monotheistic religion. Questions still swirl about him—Who really was he? What was he doing in Jerusalem? Was there a plot against him, or was he his own plotter? And what was the chalice that has given rise to the legends about (and searches for) the "Holy Grail"?

On his last evening of freedom he celebrated the Jewish Passover ceremonial meal (called *Seder* in Hebrew) with wine and unleavened bread together with his twelve disciples, and the scene has been immortalized by some of the greatest painters of religious art, Leonardo Da Vinci's *The Last Supper* being the most famous of them **(Fig. 122)**. Leonardo was renowned for his scientific knowledge and theological insights; what his painting *shows* has been

FIGURE 122

discussed, debated, and analyzed to this day—deepening, rather than resolving, the enigmas.

The key to unlocking the mysteries, we shall show, lies in what the painting *does not show*; it is what is missing from it that holds answers to troubling puzzles in the saga of God and Man on Earth, and the yearnings for Messianic Times. Past, Present, and Future do converge in the two events, separated by twenty-one centuries; Jerusalem was pivotal to both, and by their timing, they were linked by biblical prophecies about the ***End of Days***.

To understand what happened twenty-one centuries ago, we need to roll the pages of history back to Alexander, who deemed himself the son of a god, yet died in Babylon at the young age of thirty-two. While alive, he controlled his feuding generals through a mixture of favors, punishments, and even untimely deaths (some, in fact, believed that Alexander himself was poisoned). No sooner did he die than his four year-old son and his guardian, Alexander's brother, were murdered and the quarrelling generals and regional commanders divided between them the main conquered lands: Ptolemy and his successors, headquartered in Egypt, seized Alexander's African domains; Seleucus and his successors ruled, from Syria, Anatolia, Mesopotamia, and the distant Asian lands; the contested Judea (with Jerusalem) ended up in the Ptolemaic realm.

The Ptolemies, having managed to maneuver Alexander's body for burial in Egypt, considered themselves his true heirs and, by and large, continued his tolerant attitude toward others' religions. They established the famed Library of Alexandria, and assigned an Egyptian priest, known as Manetho, to write down Egypt's dynastic history and divine prehistory for the Greeks (archaeology has confirmed what is still known of Manetho's writings). That convinced the Ptolemies that their civilization was a continuation of the Egyptian one, and they thus considered themselves rightful successors to the Pharaohs. Greek savants showed particular intrerest in the religion and writings of the Jews, so much so that the Ptolemies arranged for the translation of the Hebrew Bible into Greek (a translation known as the *Septuagint*) and allowed the Jews complete religious freedom of worship in Judea, as well as in their growing communities in Egypt.

Like the Ptolemies, the Seleucids also retained a Greek-speaking scholar, a former priest of Marduk known as Berossus, to compile for them the history and prehistory of Mankind and its gods according to Mesopotamian knowledge. In a twist of history, he researched and wrote at a library of cuneiform tablets located near Harran. It is from his three books (which we know of only from fragmented quotations in the writings of others in antiquity) that the Western world, of Greece and then Rome, learnt of the Anunnaki and their coming to Earth, the prediluvial era, the creation of

Wise Man, the Deluge, and what followed. Thus it was from Berossus (as later confirmed by the discovery and decipherment of the cuneiform tablets) that the 3600 "Sar" as the "year" of the gods was first learnt.

In 200 B.C.E. the Seleucids crossed the Ptolemaic boundary and captured Judea. As in other instances, historians have searched for geopolitical and economic reasons for the war—ignoring the religious-messianic aspects. It was in the report about the Deluge that the tidbit information was given by Berossus, that Ea/Enki instructed Ziusudra (the Sumerian "Noah") to "conceal every available writing in Sippar, the city of Shamash," for post-Diluvial recovery, because those writings *were about beginnings, middles and ends*." According to Berossus, the world undergoes periodic cataclysms, and he related them to the zodiacal Ages, his contemporary one having begun 1,920 years before the Seleucid Era (312 B.C.E.); that would have placed the beginning of the Age of the Ram in 2232 B.C.E.—an Age destined to come soon to an end even if the full mathematical length is granted to it (2232–2160 = 122 B.C.E.).

The available records suggest that the Seleucid kings, coupling those calculations with the Missing Return, were seized with the need to urgently expect and prepare for one. A frenzy of rebuilding the ruined temples of Sumer and Akkad began, with emphasis on the E.ANNA—the "House of Anu"—in Uruk. The Landing Place in Lebanon, called by them Heliopolis—City of the Sun god—was rededicated by erecting a temple honoring Zeus. The reason for the war to capture Judea, one must conclude, was the urgency of also preparing the space-related site in Jerusalem for the Return. *It was, we suggest, the Greek–Seleucid way of preparing for the reappearance of the gods.*

Unlike the Ptolemies, the Seleucid rulers were determined to impose the Hellenic culture and religion in their domains. The change was most significant in Jerusalem, where suddenly foreign troops were stationed and the authority of the Temple priests was curtailed. Hellenistic culture and customs were forcefully introduced; even names had to be changed, starting with the high priest, who was obliged to change his name from Joshua to Jason. Civil laws restricted Jewish citizenship in Jerusalem; taxes were raised to finance the teaching of athletics and wrestling instead of the *Torah*; and in the countryside, shrines to Greek deities were being erected by the authorities and soldiers were sent to enforce worship in them.

In 169 B.C.E. the then Seleucid king, Antiochus IV (who adopted the epithet Epiphanes) came to Jerusalem. It was not a courtesy visit. Violating the Temple's sanctity, he entered the Holy of Holies. On his orders, the Temple's treasured golden ritual objects were confiscated, a Greek governor was put in charge of the city, and a fortress for a permanent garrison of foreign soldiers was built next to the Temple. Back in his Syrian capital, Antiochus issued a proclamation requiring worship of Greek gods throughout the kingdom; in Judea, it specifically forbade the observance of the Sabbath and circumcision. In accordance with the decree, the Jerusalem temple was to become a temple to Zeus; and in 167 B.C.E., on the 25th day of the Hebrew month Kislev—*equivalent to today's December 25*—an idol, a statue representing Zeus, "The Lord of Heaven," was set up by Syrian-Greek soldiers in the temple, and the great altar was altered and used for sacrifices to Zeus. The sacrilege could not have been greater.

The unavoidable Jewish uprising, begun and led by a priest named Matityahu and his five sons, is known as the Hashmonean or Maccabean Revolt. Starting in the countryside, the uprising quickly overcame the local Greek garrisons. As the Greeks rushed in reinforcements, the revolt engulfed the whole country; what the Maccabees lacked in numbers and weapons, they compensated for by the ferocity of their religious zeal. The events, described in the *Book of Maccabees* (and by subsequent historians), leave no doubt that the fight of the few against a powerful kingdom was guided by a certain timetable: *It was imperative to retake Jerusalem, cleanse the temple, and rededicate it to Yahweh by a certain deadline.* Managing in 164 B.C.E. to recapture only the Temple Mount, the Maccabees cleansed the Temple, and the sacred flame was rekindled that year; the final victory, leading to full control of Jerusalem and restoration of Jewish independence, took place in **160 B.C.E.** The victory and rededication of the Temple are still celebrated by Jews as the holiday of *Hanukkah* ("rededication") on the twenty-fifth day of Kislev.

The sequence and the timing of those events appeared to be linked to the prophecies about the End of Days. Of those prophecies, as we have seen, the ones that offered specific numerical clues in regard to the ultimate future, the End of Days, were conveyed by the angels to Daniel. But clarity is lacking because the counts were enigmatically expressed either in a unit called "time," or in "weeks of years," and even in numbers of days; and it is perhaps only in respect to the latter that one is told when the count does begin, so that one could know when it would end. In that one instance, the count was to begin from the day when "regular offering is abolished and an appalling abomination is set up" in the Jerusalem temple; we have established that such an abominable act indeed took place one day in 167 B.C.E.

With the sequence of those events in mind, the count of days given to Daniel must have applied to the specific events at the Temple: its defiliing in 167 B.C.E. ("when the regular offering is abolished and an appalling abomination is set up"), the cleansing of the Temple in 164 B.C.E. (after "a thousand and two hundred and ninety days"), and Jerusalem's complete liberation by 160 B.C.E. ("happy is the one who waits and reaches one thousand three hundred and thirty five days"). The numbers of days, 1290 and 1335, basically match the sequence of events at the Temple.

According to the prophecies in the *Book of Daniel,* it was then that the clock of the End of Days began ticking.

. . .

The imperative of recapturing the whole city and the removal of uncircumcised foreign soldiers from the Temple Mount by 160 B.C.E. hold the key to another clue. While we have been using the accepted count of B.C.E. and A.D. for dating events, the people of those past times obviously could not and did not use a timetable based on a *future* Christian calendar. The Hebrew calendar, as we have mentioned earlier, was the calendar begun in Nippur in 3760 B.C.E.—and according to that calendar, **what we call 160 B.C.E. was precisely the year 3600!**

That, as the reader knows by now, was a SAR, the original (mathematical) orbital period of Nibiru. And though Nibiru had reappeared four hundred years earlier, the arrival of the SAR year—3,600—**the completion of one Divine Year**—was of unavoidable significance. To those to whom the biblical prophecies of the return of Yahweh's *Kavod* to His Temple Mount were unquestioned divine pronouncements, the year we call "160 B.C.E." was a crucial moment of truth: no matter where the planet was, God has promised to Return to His Temple, and the temple had to be purified and readied for that.

That the passage of years according to the Nippurian/Hebrew calendar was not lost sight of in those turbulent times is attested by the *Book of Jubilees,* an extrabiblical book presumed to have been written in Hebrew in Jerusalem in the years following the Maccabean revolt (now available only from its Greek, Latin, Syriac, Ethiopic, and Slavonic translations). It retells the history of the Jewish people from the time of the Exodus in time units of Jubilees—the 50-year units decreed by Yahweh at Mount Sinai (see our chapter IX); it also created a consecutive calendrical historical count that has since become known as **Annu Mundi**—"Year of the World" in Latin—that starts in 3760 B.C.E. Scholars (such as the Rev. R.H. Charles in his English rendition of the book) converted such "Jubilee of years" and their "weeks" to an Anno Mundi count.

That such a calendar was not only kept throughout the ancient Near East, but even determined when events were timed to happen, can be ascertained by simply reviewing some pivotal dates (often highlighted in bold font) given in our earlier chapters. If we choose just a few of those key historical events, this is what transpires when the "B.C.E." is converted to "N.C." (Nippurian Calendar):

B.C.E.	N.C.	EVENT
3760	0	Sumerian civilization. Nipput calender begins
3460	300	The Tower of Babel incident
2860	900	Bull of Heaven killed by Gilgamesh
2360	1400	Sargon: Era of Akkad begins
2160	1600	First Intermediate Period in Egypt; Era of Ninurta (Gudea builds Temple-of-Fifty)
2060	1700	Nabu organizes Marduk's followers; Abraham to Canaan; War of the Kings
1960	**1800**	Marduk's Esagil temple in Babylon
1760	2000	Hammurabi consolidates Marduk's supremacy
1560	2200	New dynasty ("Middle Kingdom") in Egypt; new dynastic rule ("Kassite") begins in Babylon
1460	2300	Anshan, Elam, Mitanni emerge against Babylon; Moses in Sinai, the "burning bush"
960	2800	Neo-Assyrian empire launched; Akitu festival renewed in Babylon
860	2900	Ashurnasirpal wears cross symbol
760	**3000**	Prophecy in Jerusalem begins with Amos
560	3200	Anunnaki gods complete their Departure; Persians challenge Babylon; Cyrus
460	3100	Greece's golden age; Herodotus in Egypt
160	**3600**	Maccabees free Jerusalem, Temple rededicated

The impatient reader will hardly wait to fill in the next entries:

60	3700	**The Romans build the Jupiter temple at Baalbek, occupy Jerusalem**
0	3760	**Jesus of Nazareth; A.D. count begins**

The century and a half that elapsed from the Maccabean freeing of Jerusalem to the events connected with Jesus after he arrived there were some of the most turbulent in the history of the ancient world and of the Jewish People in particular.

That crucial period, whose events affect us to this day, began with understandable jubilation. For the first time in centuries the Jews were again complete masters of their holy capital and sacred Temple, free to appoint their own kings and High Priests. Though the fighting at the borders continued, the borders themselves now extended to encompass much of the olden united kingdom of David's time. The establishment of an independent Jewish state, with Jerusalem as its capital, under the Hashmoneans was a triumphal event in all respects—except one:

The return of Yahweh's *Kavod,* expected at the End of Days, did not take place, even though the count of days from abomination time seemed to have been correct. Was the Time of Fulfillment not yet at hand, many wondered; and it became evident that the enigmas of Daniel's other counts, of "years" and "weeks of years" and of "Time, Times," and so on had yet to be deciphered.

Clues were the prophetic parts in the *Book of Daniel* that spoke of the rise and fall of *future* kingdoms *after* Babylon, Persia, and Egypt—kingdoms cryptically called "of the south," "of the north," or a seafaring "Kittim"; and kingdoms that shall split off them, fight each other, "plant tabernacles of palaces between the seas"—all future entities that were also cryptically represented by varied animals (a ram, a goat, a lion and so on) whose off-spring, called "horns," will again split apart and fight each other. Who were those future nations, and what wars were foretold?

The Prophet Ezekiel also spoke of great battles to come, between north and south, between an unidentified Gog and an opposing Magog; and people were wondering whether the prophesied kingdoms have already appeared on the scene—Alexander's Greece, the Seleucids, the Ptolemies. Were these the subject of the prophecies, or was it someone yet to come in the even more distant future?

There was theological turmoil: Was the expectation at the Jerusalem Temple of the *Kavod* as a physical object a correct understanding of prophecies, or was the expected Coming only of a symbolic, of an ephemeral nature, a *spiritual Presence*? What was required of the people—or was what was destined to happen will happen no matter what? The Jewish leadership split between devout and by-the-book Pharisees and the more liberal Sadducees, who were more internationally minded, recognizing the importance of a Jewish diaspora already spread from Egypt to Anatolia to Mespotamia. In addition to these two mainstreams, small sects, sometimes organized in their own communities, sprang up; the best known of them are the Essenes (of the Dead Sea Scrolls fame), who secluded themselves at Qumran.

In the efforts to decipher the prophecies, a rising new power—**Rome**—had to be figured in. Having won repeated wars with the Phoenicians and with the Greeks, the Romans controlled the Mediterranean and began to get involved in the affairs of Ptolemian Egypt and the Seleucid Levant (Judea included). Armies followed imperial delegates; by 60 B.C.E., the Romans, under Pompey, occupied Jerusalem. On the way there, like Alexander before him, he detoured to Heliopolis (alias Baalbek) and offered sacrifices to Jupiter; it was followed by the building there, atop the earlier colossal stone blocks, of the Roman empire's greatest temple to Jupiter **(Fig. 123)**. A commemorative inscription found at the

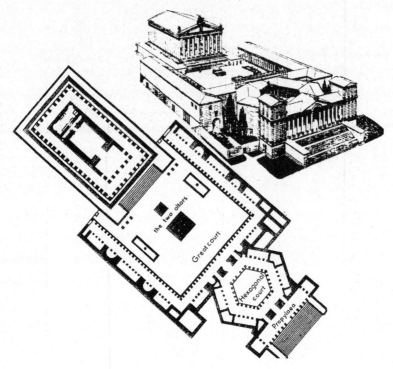

FIGURE 123

site indicates that the emperor Nero visited the place in A.D. 60, suggesting that the Roman temple was already built by then.

The national and religious turmoil of those days found expression in a proliferation of historic-prophetic writings, such as *the Book of Jubilees*, the *Book of Enoch*, the *Testaments of the Twelve Patriarchs*, and the *Assumption of Moses* (and several others, all collectively known as the *Apocrypha* and *Pseuda-Epigrapha*). The common theme in them was a belief that history is cyclical, that all has been foretold, that the End of Days—a time of turmoil and upheaval—will mark not just an end of a historic cycle but also the beginning of a new one, and that the "flipover time" (to use a modern expression) will be manifest by the coming of the **"Anointed One"**—*Mashi'ach* in Hebrew (translated *Chrystos* in Greek, and thus Messiah or Christ in English).

. . .

The act of anointing a newly invested king with priestly oil was known in the Ancient World, at least from the time of Sargon. It was recognized in the Bible as an act of consecration to God from the earliest times, but its most memorable instance was when the priest Samuel, custodian of the Ark of the Covenant, summoned David, the son of Jesse, and, proclaiming him king by the grace of God,

> *Took the horn of oil and anointed him*
> *in the presence of his brethren;*
> *and the Spirit of God*
> *came upon David from that day on.*
>   I SAMUEL 16: 13

Studying every prophecy and every prophetic utterance, the devout in Jerusalem found repeated references to **David as God's Anointed**, and a divine vow that it will be of "his seed"—by a descendant of the House of David—that his throne shall be established again in Jerusalem "*in days that are to come.*" It is on the "throne of David" that future kings, who must be of the House of David, shall sit in Jerusalem; and when that shall happen, the kings and princes of the Earth shall flock to Jerusalem for justice, peace, and the word of God. This, God vowed, is "an everlasting promise," God's covenant "for all generations." The universality of this vow is attested to in Isaiah 16: 5 and 22: 22; Jeremiah 17: 25, 23: 5, and 30: 3; Amos 9: 11; Habakuk 3: 13; Zechariah 12: 8; Psalms 18: 50, 89: 4, 132: 10, 132: 17, and so on.

These are strong words, unmistakable in their messianic covenant **with the House of David**, yet they are also full of explosive facets that virtually dictated the course of events in Jerusalem. Linked to that was the matter of the **Prophet Elijah.**

Elijah, nicknamed the Thisbite after the name of his town in the district of Gile'ad, was a biblical prophet active in the kingdom of

Israel (after the split from Judea) in the ninth century B.C.E., during the reign of king Ahab and his Canaanite wife, Queen Jezebel. True to his Hebrew name, *Eli-Yahu*—"Yahweh is my God"—he was in constant conflict with the priests and "spokesmen" of the Canaanite god Ba'al ("the Lord"), whose worship Jezebel was promoting. After a period of seclusion at a hiding place near the Jordan River, where he was ordained to become "A Man of God," he was given a "mantle of haircloth" that held magical powers, and was able to perform miracles in the name of God. His first reported miracle (I Kings Chapter 17) was the making of a spoonful of flour and a little cooking oil last a widow as food for the rest of her lifetime. He then resurrected her son, who had died of a virulent illness. During a contest with the prophets of Ba'al on Mount Carmel, he could summon a fire from the sky. His was the only biblical instance of an Israelite revisiting Mount Sinai since the Exodus: when he escaped for his life from the wrath of Jezebel and the priests of Ba'al, an Angel of the Lord sheltered him in a cave on Mount Sinai.

Of him the Scriptures said that he did not die because he was taken up to heaven in a whirlwind to be with God. His ascent, as described in great detail in II Kings Chapter 2, was neither a sudden nor an unexpected occurrence; on the contrary, it was a preplanned and prearranged operation whose place and time were communicated to Elijah in advance.

The designated place was in the Jordan Valley, on the eastern side of the river. When it was time to go there, his disciples, headed by one named Elisha, went along. He made a stop at Gilgal (where Yahweh's miracles were performed for the Israelites under the leadership of Joshua). There he tried to shake off his companions, but they went on to accompany him to Beth-El; though asked to stay put and let Elijah cross the river by himself, they stuck with him unto the last stop, Jericho, all the while asking Elisha whether it was "true that the Lord will take Elijah heavenward today?"

At the bank of the Jordan River, Elijah rolled his miracle mantle and struck the waters, parting them, enabling him to cross the river. The other disciples stayed behind, but even then Elisha persisted on being with Elijah, crossing over with him;

> And as they continued to walk and to talk,
> there appeared a chariot of fire with horses of fire,
> and the two were separated.
> And Elijah went up to heaven, in a whirlwind.
> And Elisha saw and cried out:
> "My father! My father!
> the chariot of Israel and its horsemen!"
> And he saw it no more.
>
> II KINGS 2: 11–12

Archaeological excavations at Tell Ghassul (the "Prophet's Mound"), a site in Jordan that fits the biblical tale's geography, have uncovered murals that depicted the "whirlwinds" shown in Fig. 103. It is the only site excavated under the auspices of the Vatican. (My search for the finds, which covered archaeological museums in Israel and Jordan and included a visit to the site in Jordan, and ultimately led to the Jesuit-run Pontifical Biblical Institute in Jerusalem—**Fig. 124**—is described in *The Earth Chronicles Expeditions*.)

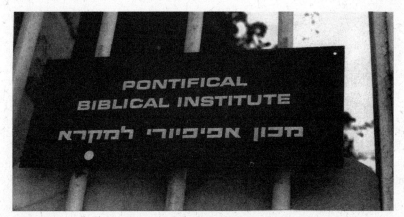

FIGURE 124

Jewish tradition has held that the transfigured Elijah will one day return as a harbinger of final redemption for the people of Israel, *a herald of the Messiah*. The tradition was already recorded in the fifth century B.C.E. by the Prophet Malachi—the last biblical Prophet—in his final prophecy. Because tradition held that the Mount Sinai cave where the angel took Elijah was where God had revealed himself to Moses, Elijah has been expected to reappear at the start of the Passover festival, when the Exodus is commemorated. To this day the *Seder*, the ceremonial evening meal when the seven-day Passover holiday begins, requires the placement on the meal table of a wine-filled cup for Elijah, to sip from as he arrives; the door is opened to enable him to enter, and a prescribed hymn is recited, expressing the hope that he will soon herald "the Messiah, son of David." (As is the case with Christian kids being told that Santa Claus did sneak down the chimney and bring them the gifts they see, so are Jewish kids told that though unseen, Elijah did sneak in and took a tiny sip of wine.) By custom, "Elijahs' Cup" has been embellished to become an artful goblet, a chalice never used for any purpose other than for the Elijah ritual at the Passover meal.

The "Last Supper" of Jesus was that tradition-filled Passover meal.

Though retaining the semblance of choosing its own high priest and king, Judea became for all intents and purposes a Roman colony, ruled first from headquarters in Syria, then by local governors. The Roman governor, called Procurator, made sure that the Jews chose as *Ethnarch* ("Head of the Jewish Council") to serve as the Temple's High Priest, and at first also a "King of the Jews" (not "King of Judea" as a country), whomever Rome preferred. From 36 to 4 B.C.E. the king was Herod, descended of Edomite converts to Judaism, who was the choice of two Roman generals (of Cleopatra fame): Mark Anthony and Octavian. Herod left a legacy of monumental structures, including the enhancement of the Temple Mount and the strategic palace-cum-fortress of Masada at the Dead Sea; he also paid heed to the governor's wishes as a de facto vassal of Rome.

It was into a Jerusalem enlarged and magnified by Hashmonean and Herodian constructions, thronged with pilgrims for the Passover holiday, that Jesus of Nazareth arrived—in A.D. 33 (according to the accepted scholarly dating). At that time the Jews were allowed to retain only a religious authority, a council of seventy elders called the *Sanhedrin*; there was no longer a Jewish king; the land, no longer a Jewish state but a Roman province, was governed by the Procurator Pontius Pilate, ensconced in the Antonia Citadel that adjoined the Temple.

Tensions between the Jewish populace and the Roman masters of the land were rising, and resulted in a series of bloody riots in Jerusalem. Pontius Pilate, arriving in Jerusalem in A.D. 26, made matters worse by bringing into the city Roman legionnaires with their pole-mounted *signae* and coinage, bearing graven images forbidden in the Temple; Jews showing resistance were pitilessly sentenced to crucifixion in such numbers that the place of execution was nicknamed *Gulgatha*—Place of the Skulls.

Jesus had been to Jerusalem before; "His parents went to Jerusalem every year at the feast of Passover, and when he was twelve years old they went up to Jerusalem after the custom of the feast; and when they had fulfilled the days, as they returned, the child Jesus tarried behind in Jerusalem" (Luke 2: 41–43). When Jesus arrived (with his disciples) this time, the situation was certainly not what was expected, not what the biblical prophecies promised. Devout Jews—as Jesus most certainly was—were beholden to the idea of redemption, of salvation by a Messiah, central to which was the special and everlasting bond between God and the House of David. It was clearly and most emphatically expressed in the magnificent Psalm 89 (19–29), in which Yahweh, speaking to His faithful followers in a vision, said:

> *I have exalted one chosen out of the people;*
> *I have found David, my servant;*
> *With my holy oil have I anointed him . . .*
> *He shall call out to me:*
> *"Thou art my father, my God,*
> *the rock of my salvation!"*
> *And I as a Firstborn shall place him,*
> *supreme of all the kings on Earth.*
> *My compassion for him forever I will keep,*
> *My faithfulness I shall not betray;*
> *My covenant with him will not be violated,*
> *What I have uttered I shall not change . . .*
> *I shall make his seed endure forever,*
> *his throne [endure]* as the **Days of Heaven.**

Was not that reference to the "Days of Heaven" a clue, a linkage between the coming of a Savior and the prophesied End of Days? Was it not the time to see the prophecies come true? And so it was that Jesus of Nazareth, now in Jerusalem with his twelve disciples, determined to take matters into his own hands: if salvation requires an Anointed One of the House of David, he, Jesus, would be the one!

His very Hebrew name—*Yehu-shuah* ("Joshua")—meant Yahweh's Savior; and as for the requirement that the Anointed One ("Messiah") be of the House of David, that he was: the very opening verse of the New Testament, in *the Gospel According to St. Matthew,* says: ***"The book of the generations of Jesus Christ, the son of David, the son of Abraham."*** Then, there and elsewhere in the New Testament, the genealogy of Jesus is given through the generations: Fourteen generations from Abraham to David; fourteen generations from David to the Babylonian exile; and fourteen generations from then to Jesus. He was qualified, the Gospels assured one and all.

Our sources for what happened next are the gospels and other books of the New Testament. We know that the "eyewitness reports" were in fact written long after the events; we know that the codified version is the result of deliberations at a convocation called by the Roman emperor Constantine three centuries later; we know that "gnostic" manuscripts, like the Nag Hammadi documents or the Gospel of Judas, give different versions that the Church had reason to suppress; we even know—which is an undisputed fact—that at first there was a Jerusalem Church led by the brother of Jesus, aimed exclusively at Jewish followers, that was overtaken, superseded, and eliminated by the Church of Rome that addressed the gentiles. Yet follow we shall the "official" version, for it, by itself, links the Jesus events in Jerusalem to all the previous centuries and millennia, as told heretofore in this book.

First, any doubt, if it still exists, that Jesus came to Jerusalem at Passover time and that the "Last Supper" was the Passover *Seder* meal must be removed. Matthew 26: 2, Mark 14: 1, and Luke 22: 1 quote Jesus saying to his disciples as they arrived in Jerusalem: "Ye know that after two days is the Feast of the Passover"; "After two days was the feast of the Passover, of the unleavened bread"; and "Now the feast of the unleavened bread drew nigh, and it is called the Passover." The three gospels, in the same chapters, then state that Jesus told his disciples to go to a certain house, where they would be able to celebrate the Passover meal with which the holiday begins.

Next to be tackled is the matter of Elijah, the herald of the coming Messiah (Luke 1: 17 even quoted the relevant verses in Malachi). According to the Gospels, the people who heard about the miracles that Jesus performed—miracles that were so similar to those by the prophet Elijah—at first wondered whether Jesus was Elijah reappeared. Not saying no, Jesus challenged his closest disciples: "What say *you* that I am? And Peter answered and said unto him: Thou art the Anointed One" (Mark 8: 28–29).

If so, he was asked, where is Elijah, who had to appear first? And Jesus answered: Yes, of course, but he has already come!

> *And they asked him, saying:*
> *Why say the scribes that Elias must first?*
> *And he answered and told them:*
> *Elias verily cometh first, and restoreth all things . . .*
> *But I say unto you*
> That Elias has indeed come.
>
> Mark 9: 11,13

This was an audacious statement, the test of which was about to come: for if Elijah has in fact returned to Earth, "*is indeed come*," thereby fulfilling the prerequisite for the Messiah's coming—**then he had to show up at the Seder and drink from his cup of wine!**

As custom and tradition required, the Cup of Elijah, filled with wine, was set on the *Seder* table of Jesus and his disciples. The ceremonial meal is described in Mark, Chapter 14. Conducting the *Seder*, Jesus took the unleavened bread (now called *Matzoh*) and made the blessing, and broke it, and gave pieces of it to his disciples. "And he took **the cup**, and when he had thanks, he gave it to them, and they all drank of it" (Mark 14: 23).

So, without doubt, the Cup of Elijah was there, but Da Vinci chose not to show it. In this *The Last Supper* painting, which could only be based on the New Testament passages, *Jesus is not holding the crucial cup, and nowhere is there a wine cup on the table!* Instead there is **an inexplicable gap to the right of Jesus (Fig. 125)**, and the disciple to his right is bending sideways as if to allow someone unseen to come between them:

FIGURE 125

Was the thoroughly theologically correct Da Vinci implying that an unseen Elijah did come through the open windows, behind Jesus, and took away the cup that was his? Elijah, the painting thereby suggests, did return; the herald preceding the Anointed King of the House of David did arrive.

And thus confirmed, when the arrested Jesus was brought before the Roman governor who asked him: "Art thou the *king of the Jews*? Jesus said unto him: Thou sayest" (Matthew 27: 11). The sentence, to die on the cross, was inevitable.

When Jesus raised the cup of wine and made the required blessing, he said to his disciples, according to Mark 14: 24, "This is my blood of the new testament." IF these were his exact words, he did not mean to say that they were to drink wine turned to blood—a grave transgression of one of the strictest prohibitions of Judaism from the earliest times, "for blood is the soul." What he said (or meant to say) was that the wine in *this cup*, the *Cup of Elijah*, was a testament, a confirmation of his *bloodline*. And Da Vinci depicted it convincingly by its disappearance, presumably taken away by the visiting Elijah.

The vanished cup has been a favorite subject of authors over the centuries. The tales became legends: the Crusaders sought it; Knights Templar found it; it was brought over to Europe . . . the cup became a goblet, a chalice; it was the chalice representing the Royal Blood—*Sang Real* in French, becoming *San Greal*, the **Holy Grail**.

Or had it, after all, never left Jerusalem?

The continued subjugation and intensified Roman repression of the Jews in Judea led to the outbreak of Rome's most challenging rebellion; it took Rome's greatest generals and best legions seven years to defeat little Judea and reach Jerusalem. In A.D. 70, after a prolonged siege and fierce hand-to-hand battles, the Romans breached the Temple's defenses; and the commanding general, Titus, ordered the Temple put to the torch. Though resistance continued elsewhere for another three years, the Jewish Great Revolt was over. The triumphant Romans were so jubilant that they commemorated the victory with a series of coins that announced to the world *Judaea Capta*—Judea Captured—and erected a victory archway in Rome depicting the looted Temple's ritual objects **(Fig. 126)**.

FIGURE 126

But during each year of independence, Jewish coins were struck with the legend "Year One," "Year Two," etc., "for the freedom of Zion," showing fruits of the land as decorative themes. **Inexplicably, the coins of years two and three bore the image of a chalice (Fig. 127) . . .**

FIGURE 127

Was the "Holy Grail" still in Jerusalem?

Chapter XVI

ARMAGEDDON AND PROPHECIES
OF THE RETURN

Will they return? When will they return?

These questions have been asked of me countless times, "they" being the Anunnaki gods whose saga has filled my books. The answer to the first question is yes; there are clues that need to be heeded, and the prophecies of the Return need to be fulfilled. The answer to the second question has preoccupied Mankind ever since the watershed events in Jerusalem more than two thousand years ago.

But the question is not only "if" and "when." What will the Return signal, what will it bring with it? Will it be a benevolent coming, or—as when the Deluge was looming—bring about the End? Which prophecies would come true: **a Messianic Time, the Second Coming, a new Beginning—or perhaps a catastrophic Apocalypse, the Ultimate End, Armageddon . . .**

It is the last possibility that shifts these prophecies from the realm of theology, escatology, or mere curiosity to a matter of Mankind's very survival; for *Armageddon,* a term that has come to denote a war of unimagined calamitous scope, *is in fact the name of a specific place in a land that has been subjected to threats of nuclear annihilation.*

In the twenty-first century B.C.E., a war of the Kings of the East against the Kings of the West was followed by a nuclear calamity. Twenty-one centuries later, when B.C.E. changed to A.D., Mankind's fears were expressed in a scroll, hidden in a cave near the Dead Sea, that described a great and final "War of the Sons of Light Against the Sons of Darkness." Now again, in the twenty-first century A.D., a nuclear threat hangs over the very same historical place. It is enough reason to ask: *Will* history repeat itself—*does* history repeat itself, in some mysterious way, every twenty-one centuries?

A war, an annihilating conflagration, has been depicted as part of the End of Days scenario in Ezekiel (chapters 38–39). Though "Gog of the land of Magog," or "Gog and Magog," are foreseen as the principal instigators in that final war, the list of combatants that shall be sucked into the battles encompassed virtually every nation of note; and the focus of the conflagration shall be "the dwellers of the Navel of the Earth"—the people of Jerusalem according to the Bible, but the people of "Babylon" as a replacement for Nippur to those for whom the clock stopped there.

It is a spine-chilling realization that Ezekiel's list of those widespread nations (38: 5) that will engage in the final war—Armageddon—actually begins with **PERSIA**—*the very country (today's Iran) whose leaders seek nuclear weapons with which to "wipe off the face of the Earth" the people who dwell where Har-Megiddo is!*

Who is that "Gog of the land of Magog," and why does that prophecy from two and a half millennia ago sound so much like current headlines? Does the accuracy of such details in the Prophecy point to the When—**to our time, to our century?**

Armageddon, a Final War of Gog and Magog, is also an essential element of the End of Days scenario of the New Testament's prophetic book, *Revelation* (whose full name is *The Apocalypse of St. John the Divine).* It compares the instigators of the apocryphal events to two beasts, one of which can "make fire come down from heaven to earth, in sight of men." Only an enigmatic clue is given for its identity (13: 18):

> *Here is wisdom:*
> *Let him that hath understanding*
> *count the number of the beast:*
> *It is the number of a man;*
> *and his number is*
> *six hundred and threescore and six.*

Many have attempted to decipher the mysterious number **666,** assuming it is a coded message pertaining to the End of Days. Because the book was written when the persecution of Christians in Rome began, the accepted interpretation is that the number was a code for the oppressor emperor Nero, the numerical value of whose name in Hebrew (NeRON QeSaR) added up to 666. The fact that he had been to the space platform in Baalbek, possibly to inaugurate the temple to Jupiter there, in the year A.D. **60** may—or may not—have a bearing on the 666 puzzle.

That there could be more to 666 than a connection to Nero is suggested by the intriguing fact that 600, 60, and 6 are all basic numbers of the Sumerian sexagesimal system, so that the "code" might hark back to some earlier texts; there were 600 Anunnaki, Anu's numerical rank was 60, Ishkur/Adad's rank was 6. Then, if the three numbers are to be multiplied rather than added, we get $666 = 600 \times 60 \times 6 = 216,000$, which is the familiar 2160 (a zodiacal age) times 100—a result that can be speculated on endlessly.

Then there is the puzzle that when seven angels reveal the sequence of future events, they do not link them to Rome; they link them to "**Babylon**." The conventional explanation has been that, like the 666 was a code for the Roman ruler, so was "Baby-lon" a code word for Rome. But Babylon was already gone for centuries when *Revelation* was written, and *Revelation,* speaking of Babylon, unmistakably links the prophecies to "the great river Euphrates" (9: 14), even describing how "the sixth angel poured out his vial upon the great river Euphrates," drying it up so that the Kings of the East would be joined in the fighting (16: 12). The talk is of a city/land on the Euphrates, not on the Tiber River.

Since *Revelation*'s prophecies are of the future, one must con-clude that ***"Babylon" is not a code—Babylon means Babylon, a future Babylon*** that will get involved in the war of "Armageddon" (which verse 16: 16 correctly explains as the name of "a place in the Hebrew tongue"—*Har-Megiddo*, Mount Megiddo, in Israel)—a war involving the Holy Land.

If that future Babylon is indeed today's Iraq, the prophetic verses are again chilling, for as they foretell current events lead-ing to the fall of Babylon after a brief but awesome war, they ***predict the breakup of Babylon/Iraq into three parts!*** (16: 19).

Like the *Book of Daniel,* which predicted phases of tribulations and trying stages in the messianic process, so has *Revelation* tried to explain the enigmatic Old Testament prophecies by describing (Chapter 20) a First Messianic Age with "a First Resurrection" lasting a thousand years, followed by a Satanic reign of a thou-sand years (when "Gog and Magog" will engage in an immense war), and then a second messianic time and another resurrection (and thus the "Second Coming").

Unavoidably, these prophecies triggered a frenzy of specula-tion as the year A.D. 2000 approached: speculation regarding the **Millennium** as a point in time, in the history of Mankind and the Earth, when prophecies would come true.

Besieged with millennium questions as the year 2000 neared, I told my audiences that ***nothing will happen in 2000***, and not only because the true millennium point counting from the birth of Jesus had already passed, Jesus having been born, by all scholarly calculations, in 6 or 7 B.C.E. The main reason for my opinion was that the prophecies appeared to envision not a *linear* timeline—year one, year two, year nine hundred, and so on—but *a cyclical* repetition of events, the fundamental belief that "The First Things shall be the Last Things"—something that can happen only when history and historical time move in a circle, where the start point is the end point, and vice versa.

Inherent in this cyclical plan of history is the concept of God as an *everlasting divine entity* who had been present at the Beginning when Heaven and Earth were created and who will be there at the End of Days, when His kingdom shall be renewed upon His holy mount. It is expressed in repeated statements from the earliest biblical assertions through the latest Prophets, as when God announced, through Isaiah (41: 4, 44: 6, 48: 12):

> *I am He, I am the First and also the Last I am . . .*
> *From the Beginnings the Ending I foretell,*
> *and from ancient times the things that are not yet done.*
>
> ISAIAH 48: 12, 46: 10

And equally so (twice) in the New Testament's *Book of Revelation*:

> *I am Alpha and Omega,*
> *the Beginning and the Ending,*
> *sayeth the Lord—*
> *Which is, and which was, and which will be.*
>
> REVELATION 1: 8

Indeed, the basis for prophecy was the belief that the End was anchored in the Beginning, that the *Future* could be predicted because the *Past* was known—if not to Man, then to God: I am the one "*who from the Beginning tells the End,*" Yahweh said (Isaiah 46: 10). The Prophet Zechariah (1: 4, 7: 7, 7: 12) foresaw

God's plans for the future—*the Last Days*—in terms of the Past, *the First Days*.

This belief, which is restated in the Psalms, in Proverbs, and in the Book of Job, was viewed as a universal divine plan for the whole Earth and all its nations. The Prophet Isaiah, envisioning the Earth's nations gathered to find out what is in store, described them asking each other: "Who among us can tell the future by letting us hear the First Things?" (41: 22). That this was a universal tenet is shown in a collection of *Assyrian Prophecies,* when the god Nabu told the Assyrian king Esarhaddon: **"The future shall be like the past."**

This cyclical element of the biblical Prophecies of the Return leads us to one current answer to the question of WHEN.

A cyclical revolving of historical time was found, the reader will recall, in Mesoamerica, resulting from the meshing, like the gears of wheels, of two calendars (see Fig. 67), creating the "bundle" of 52 years, on the occurring of which—after an unspecified number of turns—Quetzalcoatl (alias Thoth/Ningishzidda) promised to return. And that introduces us to the so-called *Mayan Prophecies,* according to which **the End of Days will come about in A.D. 2012.**

The prospect that the prophesied crucial date is almost at hand has naturally attracted much interest, and merits explaining and analyzing. The claimed date arises from the fact that in that year (depending how one calculates) the time unit called *Baktun* will complete its thirteenth turn. Since a Baktun lasts 144,000 days, it is some kind of a milestone.

Some errors, or fallacious assumptions, in this scenario need to be pointed out. The first is that the Baktun belongs not to the two "meshing" calendars with the 52-year promise (the *Haab* and the *Tzolkin*), but to a third and much older calendar called *The Long Count.* It was introduced by the Olmecs—Africans who had come to Mesoamerica when Thoth was exiled from Egypt—and the count of days actually began with that event, so that Day One of the Long Count was in what we date as August 3113 B.C.E. Glyphs in that calendar represented the following sequence of units:

1 kin			=	1 day
1 Uinal	=	1 kin × 20	=	20 days
1 Tun	=	1 kin × **360**	=	360 days
1 Ka-tun	=	1 tun × 20	=	7,200 days
1 Bak-tun	=	1 Ka-tun × 20	=	144,000 days
1 Pictun	=	1 Bak-tun × 20	=	2,880,000 days

These units, each a multiple of the previous one, thus continued beyond the Baktun with ever-increasing glyphs. But since Mayan monuments never reached beyond 12 Baktuns, whose 1,728,000 days were already beyond the Mayan existence, the 13th Baktun appears as a real milestone. Besides, Mayan lore purportedly held that the present "Sun" or Age would end with the 13th Baktun, so when its number of days (144,000 × 13 = 1,872,000) is divided by 365.25, it results in the passage of 5,125 years; when the B.C.E. 3113 is deducted, **the result is the year A.D. 2012.**

This is an exciting as well as an ominous prediction. But that date has been challenged, already a century ago, by scholars (like Fritz Buck, *El Calendario Maya en la Cultura de Tiahuanacu*) who pointed out that as the above list indicates, the mutiplier, and thus the divider, should be the calendar's own mathematically perfect 360 and not 365.25. That way, the 1,872,000 days result in 5,200 years—a perfect result, because it represents exactly 100 "bundles" of Thoth's magical number 52. Thus calculated, **Thoth's magical year of the Return would be A.D. 2087** (5200 − 3113 = 2087).

One could stand even that wait; the only fly in the ointment is that *the Long Count is a linear time count, and not the required cyclical one*, so that its counted days could roll on to the fourteenth Baktun and the fifteenth Baktun and on and on.

All that, however, does not eliminate the significance of a prophetic millennium. Since the source of "millennium" as an escatological time had its origins in Jewish apocryphal writings from the 2nd century B.C.E., the search for meaning must shift in that direction. In fact, the reference to "a thousand"—a millennium—as defining an era had its roots way back in the Old Testament. Deuteronomy (7: 9) assigned to the duration of God's covenant with Israel a period of "a thousand generations"—an assertion repeated (I Chronicles 16:15) when the Ark of the Covenant was brought to Jerusalem by David. The Psalms repeatedly applied the number "thousand" to Yahweh, his wonders, and even to his chariot (Psalm 68: 17).

Directly relevant to the issue of the End of Days and the Return is the statement in Psalm 90: 4—a statement attributed to Moses himself—that said of God that *"a thousand years, in thy eyes, are but as one day that has passed."* This statement has given rise to speculation (which started soon after the Roman destruction of the Temple) that it was a way to figure out the elusive messianic End of Days: if Creation, "The Beginning," according to *Genesis,* took God six days, and a divine day lasts a thousand years, the result is a duration of 6,000 years from Beginning to End. The End of Days, it has thus been figured, will come in the *Anno Mundi* year 6,000.

Applied to the Hebrew calendar of Nippur that began in 3760 B.C.E., **this means that the End of Days will occur in A.D. 2240** (6000 − 3760 = 2240).

This third End of Days calculation may be disappointing or comforting—it depends on one's expectations. The beauty of this calculation is that it is in perfect harmony with the Sumerian sexagesimal ("base 60") system. It might even prove in future to be correct, but I don't think so: it is again linear—and it is a cyclical time unit that is called for by the prophecies.

With none of the "modern" predicted dates workable, one must look back at the olden "formulas"—do what had been advised in Isaiah, "**look at the signs backwards.**" We have two *cyclical* choices: the Divine Time orbital period of Nibiru, and the Celestial Time of the zodiacal Precession. Which one is it?

That the Anunnaki came and went during a "window of opportunity" when Nibiru arrived at Perigee (nearest the Sun, and thus the closest to Earth and Mars) is so obvious that some readers of mine used to simply deduct 3600 from 4000 (as a round date for Anu's last visit), resulting in 400 B.C.E., or deduct 3,600 from 3760 (when the Nippur calendar began)—as the Maccabees did—and arrive at 160 B.C.E. Either way, the next arrival of Nibiru is way in the distant future.

In fact, as the reader now knows, Nibiru arrived earlier, circa 560 B.C.E. When considering that "digression," one must keep in mind that the perfect SAR (3600) has always been a *mathematical* orbital period, because celestial orbits—of planets, comets, asteroids—digress from orbit to orbit due to the gravitational tug of other planets near which they pass. To use the well-tracked Halley's Comet as an example, its given period of 75 years actually fluctuates from 74 to 76; when it last appeared in 1986, it was 76 years. Extend Halley's digression to Nibiru's 3600, and you get a plus/minus variant of about 50 years each way.

There is one other reason for wondering why Nibiru had digressed so much from its wonted SAR: the unusual occurrence of the Deluge circa 10900 B.C.E.

During its 120 SARs *before* the Deluge, Nibiru orbited without causing such a catastrophe. Then something unusual happened that brought Nibiru closer to Earth: combined with the slippage conditions of the ice sheet covering Antarctica, the Deluge occurred. What was that "something unusual"?

The answer may well lie farther out in our solar system, where Uranus and Neptune orbit—planets whose many moons include some that, inexplicably, orbit in an "opposite" ("retrograde") direction—the way Nibiru orbits.

One of the great mysteries in our solar system is the fact that the planet Uranus literally lies on its side—its north-south axis faces the Sun horizontally instead of being vertical to it. "Something" gave Uranus a "big whack" sometime in its past, NASA's scientists said—without venturing to guess what the "something" was. I have often wondered whether that "something" was also what caused the huge mysterious "chevron" scar and an unexplained "ploughed" feature that NASA's *Voyager 2* found on Uranus's moon Miranda in 1986 **(Fig. 128)**—a moon that is different in numerous ways from the other moons of Uranus. ***Could a celestial collision with a passing Nibiru and its moons cause all that?***

In recent years astronomers have ascertained that the outer large planets have not stayed put where they were formed, but have been drifting outward, away from the Sun. The studies concluded that the shift has been most pronounced in the case of

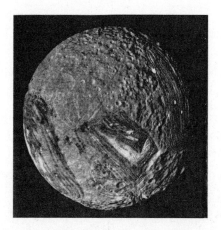

FIGURE 128

Uranus and Neptune (see sketch, **Fig. 129**), and that can explain why nothing happened out there for many Nibiru orbits—then suddeny something did. It is not implausible to assume that on its "Deluge" orbit Nibiru encountered the drifting Uranus, *and one of Nibiru's moons struck Uranus,* tilting it on its side; it could even be that the strike "weapon" was the enigmatic moon Miranda—*a moon of Nibiru*—striking Uranus and ending up captured to orbit Uranus. Such an occurrence would have affected the orbit of Nibiru, slowing it down to about 3450 Earth-years rather than 3600, and resulting in a post-Diluvial reappearance schedule of circa 7450, circa 4000, and circa 550p B.C.E.

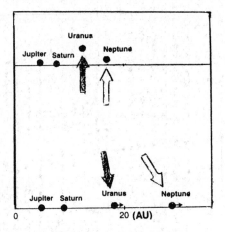

FIGURE 129

If that is what had happeed, it would explain the "early" arrival of Nibiru in 556 B.C.E.—and suggest that **its next arrival will be circa A.D. 2900**. For those who associate the prophesied cataclysmic events with the return of Nibiru—"Planet X" to some—the time is not at hand.

But any notion that the Anunnaki limited their comings and goings to a single short "window" at the planet's perigee is, however, incorrect. They could keep coming and going at other times as well.

The ancient texts record numerous instances of back-and-forth travel by the gods with no indication of a link to the planet's proximity. There are also a number of tales of Earth-Nibiru travel by Earthlings that omit any assertion of Nibiru seen in the skies (a sight stressed, on the other hand, when Anu visited Earth circa 4000 B.C.E.). In one instance Adapa, a son of Enki by an Earthling woman, who was given Wisdom but not immortality, paid a very short visit to Nibiru, accompanied by the gods Dumuzi and Ningishzidda. Enoch, emulating the Sumerian Enmeduranki, also came and went, twice, in his lifetime on Earth.

This was possible in at least two ways, as shown in **Fig. 130**: one by a spaceship accelerating on Nibiru's incoming phase (from point **A**), arriving well ahead of perigee time; the other by decel-

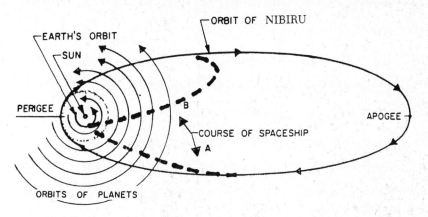

FIGURE 130

erating a spacecraft (at point **B**) during Nibiru's outbound phase, "falling back" toward the Sun (and thus to Earth and Mars). A short visit to Earth, like the one by Anu, could take place by combining "A" for arrival and "B" for outbound departure; a short visit to Nibiru (as by Adapa) could take place by reversing the procedure—by leaving Earth to intercept Nibiru at "A" and departing from Nibiru at "B" for the return to earth, and so on.

A Return of the Anunnaki at a time other than the planet's return can thus take place, and for that we are left with the other cyclical time—zodiacal time.

I have called it, in *When Time Began*, **Celestial Time**, distinct from yet serving as a link between Earthly Time (our planet's orbital cycle) and Divine Time (the clock of the Anunnaki's planet). If the expected Return will be of the Anunnaki rather than of their planet, then it behooves us to seek the solution to the enigmas of gods and men through the clock that has linked them—the cyclical zodiac of Celestial Time. It was invented, after all, by the Anunnaki as a way to reconcile the two cycles; their ratio—3600 for Nibiru, 2160 for a zodiacal Age—was the Golden Ratio of 10:6. It resulted, I have suggested, in the sexagesimal system on which Sumerian mathematics and astronomy were based (6 × 10 × 6 × 10 and so on).

Berossus, as we have mentioned, deemed the zodiacal Ages to be turning points in the affairs of gods and men and held that the world periodically undergoes apocalyptic catastrophes, either by water or by fire, whose timing is determined by heavenly phenomena. Like his counterpart Manetho in Egypt, he also divided prehistory and history into divine, semidivine, and postdivine phases, with a grand total of 2,160,000 years of "the duration of this world." This—wonder of wonders!—is **exactly one thousand—a millennium!—zodiacal ages.**

Scholars studying ancient clay tablets dealing with mathematics and astronomy were astounded to discover that the tablets used the fantastic number 12960000—yes, 12,960,000—as a starting point. They concluded that this could only be related to the zodiacal ages of 2,160, whose multiples result in 12,960 (if 2,160 × 6), or 129,600 (if 2,160 × 60), or 1,296,000 (if multiplied by 600); and—wonder of wonders!—the fantastic number with which these ancient lists begin, 12,960,000, **is a multiple of 2,160 by 6,000—as in the divine six days of creation.**

That major events, when the affairs of the gods affected the affairs of men, were linked to zodiacal ages has been shown throughout this volume of *The Earth Chronicles*. As each Age began, something momentous took place: the Age of Taurus signaled the grant of civilization to Mankind. The Age of Aries was ushered in by the nuclear upheaval and ended with the Departure. The Age of Pisces arrived with the destruction of the Temple and the beginning of Christianity. *Should one not wonder whether the prophetic End of Days really means End of (zodiacal) Age?*

Were the "time, times, and a half" of Daniel simply a terminology referring to zodiacal ages? The possibility was pondered, some three centuries ago, by none other than Sir **Isaac Newton**. Best known for his formulation of the natural laws governing celestial motions—such as planets orbiting the Sun—his interests also lay in religious thought, and he wrote lengthy treatises about the Bible and biblical prophecies. He considered the celestial motions that he formulated to be "the mechanics of God," and he strongly believed that the scientific discoveries that began with Galileo and Copernicus and were continued by him were meant to happen when they did. This led him to pay particular attention to the "mathematics of Daniel."

In March 2003 the British Broadcasting Corporation (BBC) startled the scientific and religious establishments with a program on Newton that revealed the existence of a document, handwritten by him on front and back, that calculated the End of Days according to Daniel's prophecies.

Newton wrote his numerical calculations on one side of the sheet, and his analysis of the calculations as seven "propositions" on the paper's other side. A close examination of the document—a photocopy of which I am privileged to have—reveals that the numbers that he used in the calculations include 216 and 2160 several times—a clue to me to understand what his line of thought was: *he was thinking of zodiacal time—to him, that was the Messianic Clock!*

He summed up his conclusions by writing down a set of three "not before" and a "not later than" timetable for Daniel's prophetic clues:

- Between 2132 and 2370 according to one clue given to Daniel,
- Between 2090 and 2374 according to a second clue,
- Between 2060 and 2370 for the crucial "time, times & half time."

"Sir Isaac Newton predicted the world would end in the year 2060," the BBC announced. Not exactly, perhaps—but as the table of zodiacal ages in an earlier chapter shows, he was not far off the mark in two of his "not earlier than" dates: **2060 and 2090**.

The original cherished document of the great Englishman is now kept in the Department of Manuscripts and Archives of the Jewish National and University Library—*in Jerusalem*!

A coincidence?

It was in my 1990 book *Genesis Revisited* that the "Phobos Incident"—a hushed-up event—was first publicly revealed. It concerned the loss, in 1989, of a Soviet spacecraft sent to explore Mars and its possibly hollow moonlet called Phobos.

In fact, not one but two Soviet spacecraft were lost. Named *Phobos 1* and *Phobos 2* to indicate their purpose—to probe Mars' moonlet Phobos—they were launched in 1988, to reach Mars in 1989. Though a Soviet project, it was supported by NASA and European agencies. *Phobos 1* just vanished—no details or explanation were ever publicly given. *Phobos 2* did make it to Mars, and started to send back photographs taken by two cameras—a regular one and an infrared one.

Amazingly or alarmingly, they included pictures of the shadow of a cigar-shaped object flying in the planet's skies between the Soviet craft and the surface of Mars **(Fig. 131** by the two cameras). The Soviet mission chiefs described the object

FIGURE 131

that cast the shadow as "something which some may call a flying saucer." Immediately, the spacecraft was directed to shift from Mars orbit to approach the moonlet and, from a distance of 50 yards, bombard it with laser beams. *The last picture Phobos 2 sent showed a missile coming at it from the moonlet* **(Fig. 132)**. Immediately after that, the spacecraft went into a spin and stopped transmitting—destroyed by the mysterious missile.

The "Phobos incident" remains, officially, an "unexplained accident." In fact, right thereafter, a secret commission on which all

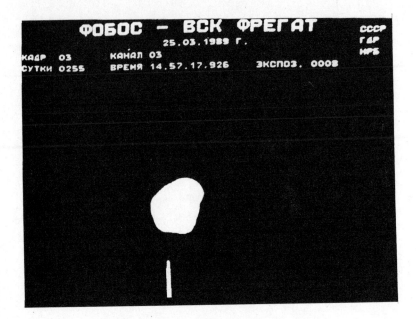

FIGURE 132

the leading space nations were represented sprang into action. The commission and the document it formulated merit more scrutiny than they received, for they hold the key to understanding what the world's leading nations really know about Nibiru and the Anunnaki.

The geopolitical events that resulted in the secret group's formation began with the discovery, in 1983, of a "Neptune-sized planet" by IRAS—NASA's Infra-Red Astronomical Satellite—which scanned the edges of the solar system not visually but by detecting heat-emitting celestial bodies. The search for a tenth planet was one of its stated objectives, and indeed it found one—determining that it was a planet because, detected once and then again six months later, it was clearly moving in our direction. The news of the discovery made headlines (Fig. 133) but was retracted the next day as a "misunderstanding." In fact, the discovery was so shocking that it led to a sudden change in U.S.–Soviet relations, a meeting and an agreement for space cooperation between President Reagan and

FIGURE 133

Chairman Gorbachev, and public statements by the President at the United Nations and other forums that included the following words (pointing heavenwards with his finger as he said them):

> **Just think how easy his task and mine might be in these meetings that we held if suddenly there was a threat to this world from some other species from another planet outside in the universe . . . I occasionally think how quickly our differences would vanish if we were facing an alien threat from outside this world.**

The Working Committee that was formed as a result of these concerns conducted several meetings and leisurely consultation—*until the March 1989 Phobos incident.* Working feverishly, it formulated in *April 1989* a set of guidelines known as the *Declaration of Principles Concerning Activities Following the Detection of Extraterrestrial*

Intelligence, by which the procedures to be followed after ***receiving "a signal or other evidence of extraterrestrial intelligence"*** were agreed upon. The "signal," the group revealed, "might not be simply one that indicates its intelligent origin but could be an ***actual message*** that may need decoding." The agreed procedures included undertakings to delay disclosure of the contact for at least twenty-four hours ***before a response is made.*** This was surely ridiculous if the message had come from a planet light years away . . . No, the preparations were for a nearby encounter!

To me, all these events since 1983, plus all the evidence from Mars briefly described in previous chapters, and the missile shot out from the moonlet Phobos, indicate that the Anunnaki still have a presence—probably a robotic presence—on Mars, their olden Way Station. That could indicate forethought, a plan to have a facility ready for a future revisit. **Put together, it suggests an *intent* for a Return.**

To me, the Earth-Mars cylinder seal (see Fig. 113) is both a depiction of the Past and a foretelling of the Future because it bears a date—*a date indicated by the sign of two fishes—the Age of Pisces.*

Does it tell us: What had taken place in a previous Age of Pisces will be repeated again in the Age of Pisces? If the prophecies are to come true, if the First Things shall be the Last Things, if the Past is the Future—the answer has to be Yes.

We are still in the Age of Pisces. The Return, the signs say, will happen before the end of our current Age.

POSTSCRIPT

In November 2005 a major archaeological discovery was made in Israel. While clearing the ground for a new structure, the remains of an ancient large building came to light. Archaeologists were

FIGURE 134

summoned to supervise careful excavation. The building turned out to be a Christian church—the oldest one ever found in the Holy Land. Inscriptions in Greek suggest it was built (or rebuilt) in the third century A.D. As the ruins were cleared, a magnificent mosaic floor came into view. In its center was a depiction of TWO FISHES—the *zodiacal sign of Pisces* **(Fig. 134)**.

What is significant about that?

The site of the discovery is Megiddo, at the foot of Mount Megiddo—*Har-Megiddo*, ARMAGEDDON.

Another coincidence?